MICROWAVE TRANSMISSION-LINE IMPEDANCE DATA

The Marconi Series
Covering Advances in Radio and Radar

Series Editor:
P. S. BRANDON
Formerly Manager, Research Division,
The Marconi Company Limited

MICROWAVE TRANSMISSION-LINE IMPEDANCE DATA

M. A. R. GUNSTON
FORMERLY CHIEF OF MATHEMATICAL PHYSICS
RESEARCH DIVISION,
THE MARCONI COMPANY LIMITED

VAN NOSTRAND REINHOLD COMPANY
LONDON

New York   Cincinnati   Toronto   Melbourne

VAN NOSTRAND REINHOLD COMPANY LTD
Windsor House, 46 Victoria Street, London, S.W.1
International Offices:
New York   Cincinnati   Toronto   Melbourne
Copyright © M. A. R. Gunston 1972
Library of Congress Catalog Card No. 77-171374
ISBN 0 442 02898 9

First published 1972

Filmset by Keyspools Ltd, Golborne, Lancs, and
Printed in Great Britain by C. Tinling and Co. Ltd,
Prescot and London

PREFACE

Since the beginnings of microwave technology some thirty or so years ago, a wide variety of structures has been proposed and/or utilized for the transmission of microwave energy. This book is concerned with that class of structures loosely referred to as "transmission lines", and constitutes a unique single-volume source of information and data on transmission-line impedances.

Hitherto, the microwave technologist, research worker, or student in need of such data has had to search the pages of text-books, technical journals, and/or unpublished or difficult-to-obtain reports. In many instances, the data required has been non-existent, inadequate, or insufficiently accurate for the purpose at hand, and it is intended that the present book will, to a large extent, rectify this situation.

The primary aim has been to reduce the time and effort involved in obtaining, for a chosen line configuration (i.e. cross-sectional shape), the dimensions requisite to realize a specified impedance.

Wherever possible, therefore, numerical data have been provided, in either tabular or graphical form (or both) and in sufficient detail and to sufficient accuracy to be of direct and immediate use to the microwave technologist. In most instances, all the relevant formulae are also quoted, supplemented by brief discussions and statements as to source, methods of derivation, and accuracy. Where such material is particularly extensive, the relevant section closes with a conclusion paragraph, in which the most accurate and/or useful formulae or results are clearly defined.

In the few instances where it has not been possible to provide numerical data, for economic or other reasons, formulae alone, or, at worst, detailed references to alternative sources of information, are quoted.

In view of the increasingly widespread availability of computer facilities to engineering and design staff, the idea of looking up data in a manual or handbook may seem somewhat old-fashioned and the value of such a book, containing extensive numerical and graphical data, may well be questioned: why not evaluate the precise data one needs by means of a simple computer program? The answer lies in the economics of the situation. The utopian state in which computer programming and run-time are cost-free has not yet been achieved, nor is it likely to be within the foreseeable future. One need, therefore, only consider the time and cost involved in preparing and testing even the simplest of programs, to realize that the initial cost of the book will already have been recouped even if the data it contains are usefully employed only once or twice.

It is expected that this book will be of greatest interest and value to the microwave technologist, and will also be a most useful source of reference for the theoretician, and for the student of electronic and microwave engineering.

Accuracy and reliability will obviously have been prime objectives in preparing this book, but if any errors have escaped the thorough checking process, the author would appreciate having them brought to his notice.

Acknowledgments

I am greatly indebted to numerous colleagues within the Research Division of The Marconi Company Ltd for their extensive assistance in compiling the tables and graphs presented in this book: in particular to Mrs E. Harper and Mr J. Weale for the provision of many of the numerical data given in Chapters 4 and 6; to Mr G. Coghill for his capable manipulation and programming of many of the formulae involving elliptic integrals; and to Mr D. Blunden and Mr B. Nicholson for the provision of the data displayed in Figs. 3.6 and 6.2 respectively. I am also indebted to the Institution of Electronic and Radio Engineers for permission to use some of their material.

Grateful thanks are also due to my wife, for the patience and forbearance displayed during the tedious task of preparing, correcting and retyping the text and most of the tables.

Finally, I am deeply indebted to The Marconi Company Ltd, both for the opportunity of gathering together the material and for permission to publish it.

M. A. R. GUNSTON
GREAT BADDOW, 1971

CONTENTS

GLOSSARY

$\varepsilon_0 =$ permittivity of free space ($8 \cdot 8552 \times 10^{-12}$ farads/metre)

$\mu_0 =$ permeability of free space ($1 \cdot 256493 \times 10^{-7}$ henry/metre)

$\mu =$ magnetic permeability of a medium

$\varepsilon =$ permittivity of a medium

$\kappa =$ dielectric constant of a medium ($\kappa = \varepsilon/\varepsilon_0$)

$\kappa' =$ dielectric constant of a different medium

$\kappa_e =$ effective dielectric constant of a combination of two or more dielectric media

$Z_0 =$ characteristic impedance of a transmission line (ohm)

$Z_{0e} =$ even-mode characteristic impedance of a pair of coupled transmission lines (ohm)

$Z_{0o} =$ odd-mode characteristic impedance of a pair of coupled transmission lines (ohm)

$C =$ capacitance per unit length of line (farads/metre)

$K(k), K'(k) =$ complete elliptic integrals of the 1st kind, of modulus k

$\ln X =$ natural or Napierian logarithm of X (i.e. $\log_e X$)

$v =$ velocity of propagation

1

INTRODUCTION

1.1 BACKGROUND

The transmission of electrical energy via wires or cables has been of importance almost since the discovery of electricity itself. The transition from direct current (d.c.) to alternating current (a.c.) usage, and thence to the utilization of higher and higher frequencies, soon necessitated the development of some form of screened transmission line in order to prevent both excessive loss of energy via radiation and resultant interference with other circuits.

These requirements led, ultimately, **to the** introduction and widespread adoption of the now almost universal coaxial cable, which is as familiar to the "average householder" (in connection with his television) as to the expert microwave engineer.

However, the rapidly increasing demands of modern technology, particularly in such areas as telecommunications, satellites, electronic warfare, and computers, have led to the introduction of many other forms of transmission lines. Some are almost as adaptable as the ubiquitous "coax.", whereas others are suitable only for certain rather specialized applications. The one common factor is that before any of these lines can usefully be incorporated into circuit designs, it is necessary to be able to calculate their impedance parameters. These are defined in Section 1.5, and the

bulk of this book consists of formulae, tabulations, and graphs of these various impedances for most of the transmission-line configurations currently used in microwave and related technologies.

As mentioned in the Preface, much of the information to be presented has hitherto been available only in articles and papers widely scattered among the technical literature, and it is hoped that the present work, by assembling, collating, and extending the data available, will constitute a comprehensive source of reference for both the microwave technologist and the microwave theoretician.

1.2 SCOPE

In order to limit the volume to a reasonable size, and because design information on microwave *waveguides* is readily available elsewhere (e.g. [1–3]) the following general restrictions have been imposed.

(*a*) Mainly, only data pertaining to TEM (**T**ransverse **E**lectro**M**agnetic) mode propagation are given and, as a necessary consequence of this,

(*b*) lossless conductors and lossless dielectrics are assumed.

Although lossless media and pure TEM mode propagation are usually tacitly assumed in microwave transmission-line design, all real physical media give rise to finite, if small, losses, and, in addition, the increasingly broader frequency bands of operation demanded by modern systems requirements mean that higher-order modes than the TEM can sometimes propagate. For this reason, the restrictions stated above are only loosely adhered to, and where appropriate, information on, and references to, losses and multi-mode propagation in transmission lines are given.

The term "microwave", used adjectivally, has various interpretations; in this text it should be read as "pertaining to that band of frequencies between 200 and 20,000 Hz". Within this band, *most* of the lines described can be used at *most* frequencies: some can be used well above and/or well below the limits quoted, if suitably constructed.

1.3 SEQUENTIAL ORDER OF DESCRIPTION

Each chapter deals with a specific interrelated family of transmission lines, and within each chapter the most logical sequence of treatment would be to start with the most general configuration and then to consider each special case thereof.

However, as might be expected, in all instances the most general case is also the most difficult to analyze: often an exact analysis is impossible, and an empirical, approximate, or numerical approach has to suffice.

In each chapter, therefore, the first case to be treated will be the most symmetric version of the appropriate configuration, since an exact analysis can usually be performed. This will be followed by a logical sequence of other special cases and, where appropriate, the most general case.

1.4 PRESENTATION OF DATA

The data are presented in three forms: formulae, tabulations, and graphs. For some transmission lines, all three are available, whereas for others, because of the complexity of the mathematical analysis involved, only graphs and/or tabulations of the data obtained by numerical techniques are given.

In general, the graphs provided can be read to an accuracy which is adequate for most microwave engineering requirements, but considerable emphasis has also been placed on the provision of detailed tables of data, at sufficiently small intervals of the parameters to enable the reader to derive, either directly or by interpolation and to virtually any desired accuracy, the values needed for a particular application.

Usually, formulae are quoted without proof, since the book is intended mainly for the design engineer—interested theoreticians can obtain details of the relevant analytical derivations from the references listed at the end of each chapter.

Rationalized MKS units are used throughout the book.

1.5 DEFINITIONS AND NOTATION

As stated previously, most of the text consists of information on transmission-line impedance parameters, and for convenient reference these are defined below.

1.5.1 Characteristic Impedance, Z_0

This parameter derives its name from the fact that, for lossless uniform TEM lines, Z_0 is a *constant* which is "characteristic" of the line concerned: it is independent of frequency, and is a function of the line cross-sectional dimensions and dielectric filling *only*.

For the reader familiar with lumped-parameter low-frequency circuits, Z_0 can be defined in terms of the series impedance per unit length of line, denoted by Z (ohm/metre), and the shunt admittance per unit length, denoted by Y (mhos/metre). These parameters are constant for a transmission line of uniform cross-section (and for the lossless lines considered here will, of course, be purely inductive and purely capacitive, respectively). The line characteristic impedance is given by

$$Z_0 = \sqrt{\frac{Z}{Y}}\ \text{ohm.} \qquad (1.5.1)$$

In terms of transmission-line concepts, Z_0 is defined as "the input impedance of an infinitely-long section of the line in question". Since an infinite (uniform) line cannot give rise to any reflection of energy, it is in effect a "perfectly-matched" line.

1.5.2 Even- and Odd-Mode Characteristic Impedances, Z_{0e} and Z_{0o}

The terms "even-mode" and "odd-mode" refer to the two possible ways in which electromagnetically-coupled twin-conductor lines can be excited. If one conductor of one line is connected to the corresponding conductor of the second line, and this connection is earthed, then the remaining two conductors can be excited either

(a) both "positive" with respect to earth, i.e. in-phase, equal-amplitude excitation: this is called "even-mode" excitation; or

(b) one "positive", one "negative" with respect to earth, i.e. opposite-phase, equal-amplitude excitation: this is called "odd-mode" excitation.

These modes of excitation are illustrated schematically in Fig. 1.1, for the case of the simple wire-above-ground type of line (described in Section 2.4).

The characteristic impedances, Z_{0e} and Z_{0o}, associated with these modes of excitation, are defined in a manner similar to that for Z_0; namely, as the input impedance of an infinite length of one line, in the presence of (and thus electro-magnetically coupled to) the second line, also of infinite length, when both are excited in the appropriate manner ((a) or (b) above).

A knowledge of Z_{0e} and Z_{0o} as functions of the line parameters is essential to the design of filters, directional couplers, and related devices.

All three of the characteristic impedances defined above (Z_0, Z_{0e}, and Z_{0o}) can be simply expressed in terms of the capacitance per unit length of the particular transmission line in question: if this parameter is denoted by C (farads per metre (F/m)), then

$$Z_0\sqrt{\kappa} = \frac{1}{v_0 C} \text{ ohm} \qquad (1.5.2)$$

where v_0 is the velocity of light *in vacuo* ($2 \cdot 997925 \times 10^8$ m/s) and κ is the dielectric constant of the medium filling the line. Z_{0e} and Z_{0o} are given by

identical expressions, in which C is the capacitance appropriate to even- or odd-mode voltage excitation.

For a true TEM mode propagating line, in the absence of losses, the capacitance C is identical with the electrostatic capacitance of the structure, and hence the real physical three-dimensional problem can be reduced to the simpler two-dimensional mathematical problem of calculating the electrostatic capacitance of the line cross-section. This can frequently be solved by the use of conformal transformation techniques, in particular by means of the Schwarz–Christoffel transformation [4], as will be seen later. In other cases, however, a closed-form exact solution cannot be obtained and in such instances, approximational, semi-empirical, or numerical techniques are used to derive the required results.

1.5.3 Derivation and Use of Numerical Constants

Throughout this book, the reader will notice that many of the impedance formulae quoted involve numerical constants, e.g. $59 \cdot 952$, $376 \cdot 687$, etc. The precise values of these constants depend upon the values assumed for the "fundamental" constants ε_0 and μ_0, from which they are derived. In this text the author has chosen ε_0 (permittivity of free space) and c (velocity of light *in vacuo*) as fundamental constants, and at the time of writing the most accurate values of these constants were believed to be:

$$c = 2 \cdot 997925 \times 10^8 \text{ m/s}$$

$$\varepsilon_0 = 8 \cdot 8552 \times 10^{-12} \text{ F/m}.$$

However, it has since been noted that this value of ε_0 differs from that specified in the SI system of units, namely:

$$\varepsilon_0 = 8 \cdot 85416 \times 10^{-12} \text{ F/m}.$$

It is possible that the reader may wish to adjust the data given in this book, to conform to the SI system, and therefore the most frequently used "impedance constants" are listed below and defined in terms of their "fundamental" constants.

As far as the majority of users of this text is concerned, this difference in value of approximately $0 \cdot 01$ per cent is of purely academic importance: for all practical purposes the final choice of any numerical constant is decided by the purely

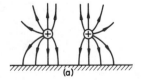
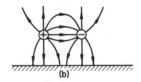

Fig. 1.1. Schematic representation of even-mode and odd-mode voltage excitation for a typical pair of coupled transmission lines, indicating the approximate form of the electric field lines and the polarity of the various conductors involved. (a) Even-mode excitation, (b) odd-mode excitation

empirical method of comparing theoretical with measured results; and differences in impedance of less than 0·1 per cent are completely insignificant for all but the most exacting requirements.

List of numerical constants used in this text

Velocity of light *in vacuo* $= c = 2 \cdot 997925 \times 10^8$ m/s.
Permittivity of free space $= \varepsilon_0 = 8 \cdot 8552 \times 10^{-12}$ F/m.
The "permeability of free space", μ_0, is defined by, and calculated from, the following relationship

$$c^2 = \frac{1}{\mu_0 \varepsilon_0}.$$

The "derived" impedance constants are as follows.

Characteristic impedance of free space $\quad Z = \sqrt{\dfrac{\mu_0}{\varepsilon_0}}$

$$\sqrt{\frac{\mu_0}{\varepsilon_0}} = 376 \cdot 687 \text{ ohm} \qquad \frac{\pi}{4}\sqrt{\frac{\mu_0}{\varepsilon_0}} = 296 \cdot 1 \text{ ohm}$$

$$\frac{1}{2}\sqrt{\frac{\mu_0}{\varepsilon_0}} = 188 \cdot 344 \text{ ohm} \qquad \frac{1}{\pi}\sqrt{\frac{\mu_0}{\varepsilon_0}} = 119 \cdot 904 \text{ ohm}$$

$$\frac{1}{4}\sqrt{\frac{\mu_0}{\varepsilon_0}} = 94 \cdot 172 \text{ ohm} \qquad \frac{1}{2\pi}\sqrt{\frac{\mu_0}{\varepsilon_0}} = 59 \cdot 952 \text{ ohm}$$

$$\frac{1}{8}\sqrt{\frac{\mu_0}{\varepsilon_0}} = 47 \cdot 086 \text{ ohm} \qquad \frac{1}{4\pi}\sqrt{\frac{\mu_0}{\varepsilon_0}} = 29 \cdot 976 \text{ ohm}$$

REFERENCES

1. Southworth, G. C. *Principles and Applications of Waveguide Transmission*.
   Van Nostrand, Princeton (1950).
2. Marcuvitz, N. "Waveguide Handbook", *M.I.T. Radiation Laboratory Series*.
   Vol. 10, McGraw-Hill, New York (1951).
3. Booth, A. E. *Microwave Data Tables*.
   Iliffe, Lond. (1959).
4. Cunningham, J. *Complex Variable Methods*.
   Van Nostrand, Lond. (1965).

2

TRANSMISSION LINES UTILIZING CONDUCTORS OF CIRCULAR CROSS-SECTION

2.1 INTRODUCTION

Two of the four transmission lines treated in this chapter have few microwave applications because of excessive losses occasioned by radiation: their presence in a book devoted to microwave topics may therefore be questioned. However, they have been included (see Sections 2.4 and 2.5) because of the close "family relationship" which exists between all four lines described, and also because the two "interlopers" are direct ancestors of the microstrip transmission line (see Section 3.6), which is now of such great importance in microwave miniaturization and integrated circuit applications.

Strictly speaking, the elliptic coaxial line should be included in this chapter, since it is the basic form from which the coaxial line and its relatives are derived. However, since its conductors are not circular in cross-section, and since it is of little practical importance, it has been included in Chapter 5, under the heading of "lines of unusual cross-section".

2.2 THE CIRCULAR COAXIAL LINE

Following the sequential order of description outlined in Section 1.3, the most symmetrical configuration of conductors is treated first.

As is implied by the name, the two cylindrical conductors which constitute this line are located coaxially, one within the other, as shown in cross-section in Fig. 2.1. The symbolism used here for the dimensional parameters departs from that in common use (in which the outer diameter is $2b$, the inner diameter, $2a$); this is simply for the sake of consistency with the notation used in the remainder of the book.

As well as displaying one of the simplest physical configurations, the coaxial line is also one of the simplest lines to analyze, either on the basis of lumped-parameter circuit concepts [1], or using a field-equation approach based upon Maxwell's Equations [2].

Exact analysis of the lossless coaxial line by

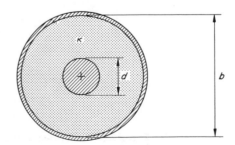

Fig. 2.1. Cross-section of circular coaxial line, filled with a dielectric medium of dielectric constant κ

either method yields the very simple expression for the characteristic impedance:

$$Z_0 = \frac{1}{2\pi}\sqrt{\frac{\mu}{\varepsilon}}\ln(b/d) \text{ ohm} \qquad (2.2.1)$$

where μ and ε are, respectively, the magnetic permeability (H/m) and dielectric permittivity (F/m) of the insulating medium filling the space between the conductors. Substituting the appropriate numerical values of the physical constants into equation (2.2.1) results in the following expression

$$Z_0\sqrt{\kappa} = 59\cdot952\ln(b/d) \text{ ohm} \qquad (2.2.2)$$

Here, κ is the dielectric constant of the insulating medium and acts as a convenient normalization parameter, thereby rendering it necessary to present only a single graph (Fig. 2.2) and tabulation (Table 2.1) in order to be able to calculate the characteristic impedance of any coaxial line with any dielectric filling. (A similar, but reciprocal (i.e. $Z_0\sqrt{\kappa}$ versus b/d), tabulation is given on p. 31 of [7].)

In making practical use of the data given, one must bear in mind the restrictions stated in

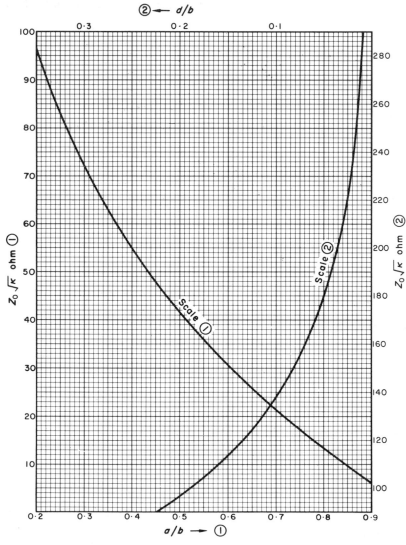

Fig. 2.2. Characteristic impedance of circular coaxial line as a function of the line parameters. Z_0 **is the characteristic impedance (ohms),** b **is the inner diameter of outer conductor,** d **is the diameter of centre conductor,** κ **is the dielectric constant of medium filling the line**

TABLE 2.1 CHARACTERISTIC IMPEDANCE OF CIRCULAR COAXIAL LINE AS A FUNCTION OF THE LINE PARAMETERS

d/b	$Z_0\sqrt{\kappa}$ (ohm)	d/b	$Z_0\sqrt{\kappa}$ (ohm)	d/b	$Z_0\sqrt{\kappa}$ (ohm)	d/b	$Z_0\sqrt{\kappa}$ (ohm)	d/b	$Z_0\sqrt{\kappa}$ (ohm)	d/b	$Z_0\sqrt{\kappa}$ (ohm)
0·010	276·09	0·210	93·56	0·410	53·45	0·610	29·63	0·810	12·63		
0·015	251·78	0·215	92·15	0·415	52·73	0·615	29·14	0·815	12·26		
0·020	234·53	0·220	90·77	0·420	52·01	0·620	28·66	0·820	11·90		
0·025	221·16	0·225	89·43	0·425	51·30	0·625	28·18	0·825	11·53		
0·030	210·23	0·230	88·11	0·430	50·60	0·630	27·70	0·830	11·17		
0·035	200·98	0·235	86·82	0·435	49·90	0·635	27·23	0·835	10·81		
0·040	192·98	0·240	85·56	0·440	49·22	0·640	26·76	0·840	10·45		
0·045	185·92	0·245	84·32	0·445	48·54	0·645	26·29	0·845	10·10		
0·050	179·60	0·250	83·11	0·450	47·87	0·650	25·83	0·850	9·74		
0·055	173·89	0·255	81·92	0·455	47·21	0·655	25·37	0·855	9·39		
0·060	168·67	0·260	80·76	0·460	46·55	0·660	24·91	0·860	9·04		
0·065	163·87	0·265	79·62	0·465	45·91	0·665	24·46	0·865	8·69		
0·070	159·43	0·270	78·50	0·470	45·27	0·670	24·01	0·870	8·35		
0·075	155·29	0·275	77·40	0·475	44·63	0·675	23·56	0·875	8·01		
0·080	151·42	0·280	76·32	0·480	44·00	0·680	23·12	0·880	7·66		
0·085	147·79	0·285	75·26	0·485	43·38	0·685	22·68	0·885	7·32		
0·090	144·36	0·290	74·21	0·490	42·77	0·690	22·25	0·890	6·99		
0·095	141·12	0·295	73·19	0·495	42·16	0·695	21·81	0·895	6·65		
0·100	138·04	0·300	72·18	0·500	41·56	0·700	21·38	0·900	6·32		
0·105	135·12	0·305	71·19	0·505	40·96	0·705	20·96	0·905	5·98		
0·110	132·33	0·310	70·21	0·510	40·37	0·710	20·53	0·910	5·65		
0·115	129·67	0·315	69·26	0·515	39·78	0·715	20·11	0·915	5·33		
0·120	127·11	0·320	68·31	0·520	39·20	0·720	19·69	0·920	5·00		
0·125	124·67	0·325	67·38	0·525	38·63	0·725	19·28	0·925	4·67		
0·130	122·32	0·330	66·47	0·530	38·06	0·730	18·87	0·930	4·35		
0·135	120·05	0·335	65·56	0·535	37·50	0·735	18·46	0·935	4·03		
0·140	117·87	0·340	64·68	0·540	36·94	0·740	18·05	0·940	3·71		
0·145	115·77	0·345	63·80	0·545	36·39	0·745	17·65	0·945	3·39		
0·150	113·74	0·350	62·94	0·550	35·84	0·750	17·25	0·950	3·08		
0·155	111·77	0·355	62·09	0·555	35·30	0·755	16·85	0·955	2·76		
0·160	109·87	0·360	61·25	0·560	34·76	0·760	16·45	0·960	2·45		
0·165	108·02	0·365	60·42	0·565	34·23	0·765	16·06	0·965	2·14		
0·170	106·23	0·370	59·61	0·570	33·70	0·770	15·67	0·970	1·83		
0·175	104·49	0·375	58·80	0·575	33·18	0·775	15·28	0·975	1·52		
0·180	102·81	0·380	58·01	0·580	32·66	0·780	14·90	0·980	1·21		
0·185	101·16	0·385	57·22	0·585	32·14	0·785	14·51	0·985	0·91		
0·190	99·56	0·390	56·45	0·590	31·63	0·790	14·13	0·990	0·60		
0·195	98·01	0·395	55·69	0·595	31·13	0·795	13·75	0·995	0·30		
0·200	96·49	0·400	54·93	0·600	30·63	0·800	13·38	1·000	0·00		
0·205	95·01	0·405	54·19	0·605	30·13	0·805	13·00				

Z_0 is the characteristic impedance, d is the diameter of centre conductor, b is the inner diameter of outer conductor, and κ is the dielectric constant of the medium filling the line interior.

Chapter 1: in particular that only TEM mode propagation is assumed. In coaxial line, the "cut-off" wavelength λ_c of the first of the possible higher-order ("waveguide") modes which can propagate is given approximately by

$$\lambda_c = \frac{\pi b}{2}\left(1+\frac{d}{b}\right). \qquad (2.2.3)$$

Hence, pure TEM mode propagation can only be assured if the coaxial line dimensions and the operating frequency are chosen to be such that the propagation wavelength λ obeys the following inequality

$$\lambda > \frac{\pi b}{2}\left(1+\frac{d}{b}\right). \qquad (2.2.4)$$

An exact general analysis of the coaxial structure has been given by Poincelot [3, 4], using the field-equation approach, and this can be used to calculate the attenuation of lines involving lossy conductors and/or dielectrics: see also [13]. However, some very useful approximate formulae for coaxial line losses, in terms of a series resistance R ohm per unit length and a shunt conductance G mhos per unit length, are quoted by Jackson (see [1], pp. 38–40), as follows

$$R = \sqrt{\frac{f}{\pi}}\left[\frac{1}{b}\sqrt{\frac{\mu_b}{\sigma_b}}+\frac{1}{d}\sqrt{\frac{\mu_d}{\sigma_d}}\right] \text{ohm/metre} \qquad (2.2.5)$$

$$G = \frac{2\pi}{1\cdot8\times10^{10}}\cdot\frac{\kappa f\tan\delta}{\ln(b/d)} \text{ mho/metre}. \qquad (2.2.6)$$

In these formulae, f is the appropriate operating frequency (in Hz), μ is the magnetic permeability and σ is the electrical conductivity of the conductor labelled by the subscript "a" or "b", and δ is the loss angle of the insulating medium.

2.3 THE ECCENTRIC COAXIAL LINE

In this version of coaxial line, the inner conductor is laterally displaced from its "normal" coaxial position, to the off-axis location shown in the cross-sectional view in Fig. 2.3. Although of little practical use for the transmission of energy, because of the difficulties of achieving accuracy and reproducibility of construction, the eccentric line is of some importance in microwave technology as an impedance transformer [5], usually

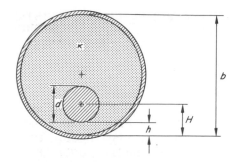

Fig. 2.3. Cross-section of eccentric coaxial line filled with a dielectric medium of dielectric constant κ

from coaxial line to one of the other types of line. For such applications, a non-uniform version is used, in which H varies with distance along the line—usually, but not necessarily, in a linear manner.

A knowledge of the parameters of eccentric coaxial line is more frequently required in the practical estimation of the magnitude of impedance mismatches which may arise from the use of imperfectly-constructed cylindrical coaxial line. The centre conductor may accidentally have become displaced, either during inadequately-controlled manufacturing processes, or as a result of excessive distortion during later handling.

Even high-quality line must obviously have a finite, if small, tolerance on conductor centralization, and in certain specialized applications, the resultant slight change in impedance may be of importance.

The eccentric line also provides a link between the coaxial line and the two "interlopers" already mentioned (which are treated in Sections 2.4 and 2.5) which themselves form direct links with the microstrip line.

The characteristic impedance of the eccentric coaxial line can be determined by quite simple field-equation analysis [6], and the resulting expression is [5–7]

$$Z_0\sqrt{\kappa} = 59\cdot952\ln(X+\sqrt{X^2-1}) \text{ ohm} \qquad (2.3.1)$$

where

$$X = \frac{1}{2}\left\{\frac{d}{b}+\frac{4H}{d}\left(1-\frac{H}{b}\right)\right\} \qquad (2.3.2)$$

and the dimensional parameters are as defined in Fig. 2.3.

In using these expressions for the calculation

of numerical data, it is convenient to express the displacement of the inner conductor from the "normal" central position in terms of a "percentage eccentricity", which will be denoted by E. This parameter expresses the distance between conductor centres (namely $b/2 - H$) as a percentage of the outer conductor radius, $b/2$. E is therefore defined as follows:

$$E = \frac{100}{b/2}\left(\frac{b}{2} - H\right) \text{per cent}$$

$$= 100\left(1 - \frac{2H}{b}\right) \text{per cent.} \tag{2.3.3}$$

Equation (2.3.2) can now be rewritten in terms of E, thus

$$X = \frac{1}{2}\left\{\frac{d}{b} + \frac{b}{d}\left(1 - \frac{E^2}{10^4}\right)\right\}. \tag{2.3.4}$$

The data tabulated in Table 2.2 and displayed graphically in Figs. 2.4 (a) and (b) were calculated from equation (2.3.1), using values of X from equation (2.3.4). Similar graphs, of less precision, can be found in [8] and [9].

Some useful information on the field patterns and cut-off frequencies of higher-order (TE and

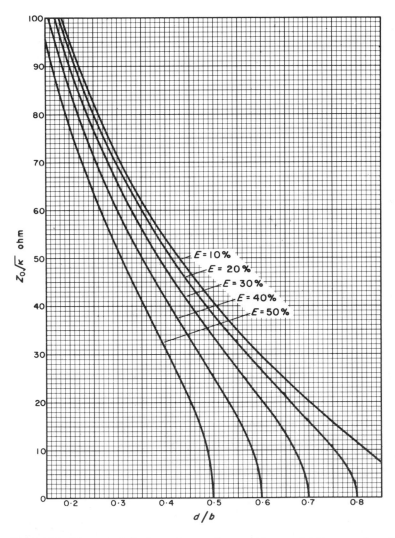

Fig. 2.4(a). Characteristic impedance of eccentric coaxial line as a function of the line parameters

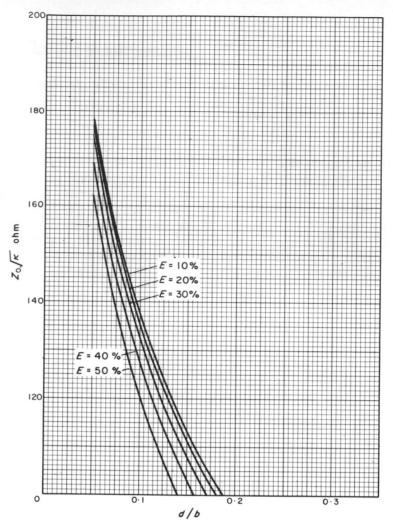

Fig. 2.4(b). Characteristic impedance of eccentric coaxial line as a function of the line parameters

TM) modes in eccentric coaxial line are given in [10] and [14], which also list further useful references on the topic.

2.4 THE SINGLE-WIRE-ABOVE-GROUND LINE

2.4.1 Introduction

The single-wire-above-ground line (the first of the two "interlopers" referred to in Section 2.1) is, in its original simple form shown in Fig. 2.5, of little use for the transmission of microwave energy, due to the high radiation loss already mentioned. This arises largely because of the unscreened nature of the line configuration.

However, matters are considerably improved (in principle at least; there still remains the practical problem of mounting a circular rod conductor upon a plane surface) if a plane slab of low-loss insulating material, having a high dielectric constant κ' ($\kappa' \gg \kappa$), is interposed between wire and ground-plane as shown in Fig. 2.6. This is because the high dielectric constant has the effect of partially concentrating the electric field within the dielectric slab, thus reducing radiation loss. In this modified form, the structure

is almost identical with that of the now-popular microstrip, described in Section 3.6, and by analogy it may be termed "round microstrip", for easy reference.

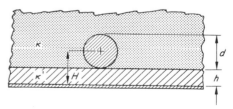

Fig. 2.5. Cross-section of the "single-wire-above-ground" line, embedded in a uniform medium of dielectric constant κ

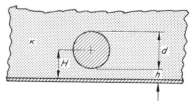

Fig. 2.6. Cross-section of the "round microstrip" version of the "single-wire-above-ground": the wire is supported upon a plane layer (substrate) of a dielectric medium of dielectric constant κ', and is surrounded by a second medium of dielectric constant κ: in general $\kappa' > \kappa$

2.4.2 The Single-Wire-above-Ground

In its simplest form (Fig. 2.5), the single-wire-above-ground can readily be analyzed by a variety of methods. For the determination of its characteristic impedance, the simplest approach is possibly via the eccentric coaxial line previously described. From Fig. 2.3 it is easily seen that if "b" becomes large, and eventually infinite, while keeping "d" and "H" constant, the end result is in fact the single-wire transmission line at present under discussion. Carrying out this process, it can be seen from equation (2.3.2) that

$$\lim_{b \to \infty} X = \frac{2H}{d} \qquad (2.4.1)$$

and hence, by substituting this result in equation (2.3.1), the characteristic impedance of the single-wire-above-ground is obtained as

$$Z_0\sqrt{\kappa} = 59 \cdot 952 \ln\left\{\frac{2H}{d} + \sqrt{\frac{4H^2}{d^2} - 1}\right\} \text{ohm} \qquad (2.4.2a)$$

where κ is the dielectric constant of the medium in which the wire is embedded. For purposes of comparison with the microstrip line described later (Section 3.6) equation (2.4.2a) can be expressed in terms of h, thus

$$Z_0\sqrt{\kappa} = 59 \cdot 952 \ln\left\{\left(1 + \frac{2h}{d}\right) + 2\sqrt{\frac{h}{d}\left(1 + \frac{h}{d}\right)}\right\} \text{ohm} \qquad (2.4.2b)$$

The limiting case for small diameter wire is of interest: for $d \ll h$ there results

$$Z_0\sqrt{\kappa} \doteq 59 \cdot 952 \ln\left(\frac{4h}{d}\right) \text{ohm.} \qquad (2.4.3)$$

The data tabulated in Table 2.3 and displayed graphically on Figs. 2.7 (a) and (b) (the curve for $\kappa'/\kappa = 1 \cdot 0$) were calculated from equation (2.4.2a).

2.4.3 "Round Microstrip"

The "round microstrip" form of the single-wire-above-ground, shown in Fig. 2.6, is much less simple to analyse. Strictly speaking, it is not a TEM mode line at all, since the boundary conditions at the interface between the two dielectrics cannot be satisfied by an electromagnetic field which consists only of transverse field components.

However, the actual field structure resembles that of the assumed TEM mode sufficiently closely for an analysis to be possible on the basis of this assumption. Such an analysis has been given by Kaden [11]: the results are, of course, rather more complex than in the single-dielectric case previously considered. Following the same procedure as Kaden, making use of the results obtained in equation (2.4.2b), and evaluating the infinite integral involved in the analysis by reference to [12], the line capacitance per unit length (C') is obtained as follows:

$$\frac{1}{C'} = \frac{1}{2\pi\kappa}\left[\ln\left\{1 + \frac{1}{2x}(1 + \sqrt{1+4x})\right\} + \sum_{n=0}^{\infty} (-D)^{n+1} \ln\left\{\frac{(n+2+x)}{(n+x)}\right\}\right] \qquad (2.4.4)$$

where $x = d/4h$, and D is a function of the ratio between the two dielectric constants:

$$D = \frac{1 - (\kappa/\kappa')}{1 + (\kappa/\kappa')} \qquad (2.4.5)$$

The line capacitance per unit length in the case of a single uniform dielectric is given by

$$\frac{1}{C} = \frac{1}{2\pi\kappa}\ln\left\{1 + \frac{1}{2x}(1 + \sqrt{1+4x})\right\} \qquad (2.4.6)$$

It is therefore possible to define an "equivalent" or "effective" dielectric constant, κ_e, for the two-dielectric configuration, as follows:

$$\kappa_e = \kappa C'/C \qquad (2.4.7)$$

and hence the velocity of propagation of the pseudo-TEM mode will be

$$v = \frac{1}{\sqrt{\mu\varepsilon_0 \kappa_e}} \text{ metres/second} \qquad (2.4.8)$$

Then, from equation (1.5.2), using equations (2.4.7) and (2.4.8), the characteristic impedance of the "round microstrip" is given by

$$Z_0\sqrt{\kappa} = 59 \cdot 952\sqrt{L(L+S)} \text{ ohm} \qquad (2.4.9)$$

where

$$L = \ln\left\{1 + \frac{1}{2x}(1 + \sqrt{1+4x})\right\} \qquad (2.4.10)$$

$$x = d/4h \qquad (2.4.11)$$

TABLE 2.2 CHARACTERISTIC IMPEDANCE OF ECCENTRIC COAXIAL LINE AS A FUNCTION OF THE ECCENTRICITY AND OTHER LINE PARAMETERS

$\frac{E}{H/b}$ d/b	5% 0·475 $Z_0\sqrt{\kappa}$	10% 0·450 $Z_0\sqrt{\kappa}$	15% 0·425 $Z_0\sqrt{\kappa}$	20% 0·400 $Z_0\sqrt{\kappa}$	25% 0·375 $Z_0\sqrt{\kappa}$	$\frac{E}{H/b}$ d/b	30% 0·350 $Z_0\sqrt{\kappa}$	35% 0·325 $Z_0\sqrt{\kappa}$	40% 0·300 $Z_0\sqrt{\kappa}$	45% 0·275 $Z_0\sqrt{\kappa}$	50% 0·250 $Z_0\sqrt{\kappa}$
0·05	179·45	179·00	178·23	177·15	175·72	0·05	173·93	171·74	169·11	165·99	162·29
0·06	168·52	168·06	167·30	166·21	164·78	0·06	162·99	160·80	158·17	155·03	151·33
0·07	159·28	158·82	158·06	156·97	155·54	0·07	153·74	151·55	148·91	145·77	142·05
0·08	151·27	150·82	150·05	148·96	147·53	0·08	145·73	143·53	140·88	137·73	134·00
0·09	144·21	143·75	142·99	141·89	140·46	0·09	138·65	136·45	133·80	130·64	126·89
0·10	137·89	137·44	136·67	135·57	134·13	0·10	132·32	130·11	127·45	124·28	120·53
0·11	132·18	131·72	130·95	129·85	128·41	0·11	126·60	124·38	121·71	118·53	114·75
0·12	126·96	126·50	125·73	124·63	123·18	0·12	121·36	119·14	116·46	113·27	109·47
0·13	122·16	121·70	120·93	119·82	118·37	0·13	116·55	114·32	111·63	108·42	104·60
0·14	117·72	117·26	116·48	115·37	113·92	0·14	112·09	109·85	107·14	103·92	100·08
0·15	113·58	113·12	112·34	111·23	109·77	0·15	107·93	105·68	102·97	99·72	95·86
0·16	109·71	109·25	108·47	107·35	105·89	0·16	104·04	101·78	99·05	95·79	91·90
0·17	106·08	105·61	104·83	103·71	102·24	0·17	100·38	98·11	95·37	92·09	88·17
0·18	102·65	102·18	101·39	100·27	98·79	0·18	96·93	94·65	91·89	88·58	84·63
0·19	99·41	98·94	98·15	97·02	95·53	0·19	93·66	91·37	88·59	85·26	81·28
0·20	96·33	95·86	95·07	93·93	92·44	0·20	90·56	88·25	85·46	82·10	78·08
0·21	93·41	92·93	92·13	91·00	89·50	0·21	87·61	85·28	82·47	79·09	75·03
0·22	90·62	90·14	89·34	88·19	86·69	0·22	84·78	82·45	79·61	76·20	72·10
0·23	87·95	87·47	86·67	85·52	84·00	0·23	82·09	79·73	76·87	73·43	69·28
0·24	85·40	84·92	84·11	82·95	81·43	0·24	79·50	77·13	74·25	70·77	66·57
0·25	82·95	82·47	81·65	80·49	78·95	0·25	77·01	74·62	71·72	68·20	63·95
0·26	80·60	80·11	79·29	78·12	76·58	0·26	74·62	72·21	69·28	65·73	61·42
0·27	78·34	77·85	77·02	75·84	74·29	0·27	72·32	69·89	66·93	63·33	58·96
0·28	76·15	75·66	74·83	73·65	72·08	0·28	70·10	67·64	64·65	61·01	56·57
0·29	74·05	73·55	72·72	71·52	69·95	0·29	67·95	65·47	62·44	58·76	54·25
0·30	72·02	71·52	70·68	69·47	67·88	0·30	65·86	63·36	60·30	56·57	51·97
0·31	70·05	69·55	68·70	67·49	65·88	0·31	63·85	61·32	58·22	54·43	49·76
0·32	68·14	67·64	66·78	65·56	63·95	0·32	61·89	59·34	56·20	52·35	47·58
0·33	66·30	65·79	64·93	63·70	62·06	0·33	59·99	57·41	54·23	50·31	45·45
0·34	64·51	63·99	63·13	61·88	60·24	0·34	58·14	55·53	52·30	48·32	43·34
0·35	62·77	62·25	61·38	60·12	58·46	0·35	56·34	53·69	50·42	46·37	41·27
0·36	61·08	60·56	59·67	58·41	56·73	0·36	54·58	51·90	48·50	44·45	39·22
0·37	59·43	58·91	58·02	56·74	55·04	0·37	52·87	50·15	46·78	42·56	37·18
0·38	57·83	57·30	56·40	55·11	53·40	0·38	51·20	48·44	45·01	40·69	35·15
0·39	56·27	55·74	54·83	53·53	51·79	0·39	49·56	46·77	43·27	38·85	33·12
0·40	54·75	54·21	53·30	51·98	50·22	0·40	47·97	45·12	41·56	37·03	31·09
0·41	53·27	52·73	51·80	50·47	48·69	0·41	46·40	43·51	39·87	35·22	29·04
0·42	51·83	51·27	50·34	48·99	47·19	0·42	44·87	41·92	38·20	33·42	26·96
0·43	50·41	49·86	48·91	47·54	45·72	0·43	43·36	40·36	36·56	31·62	24·84
0·44	49·03	48·47	47·51	46·13	44·28	0·44	41·88	38·83	34·93	29·82	22·66
0·45	47·68	47·11	46·14	44·74	42·86	0·45	40·43	37·31	33·31	28·01	20·38
0·46	46·36	45·79	44·80	43·39	41·48	0·46	39·00	35·81	31·70	26·18	17·96
0·47	45·07	44·49	43·49	42·05	40·12	0·47	37·59	34·33	30·09	24·32	15·33
0·48	43·81	43·22	42·21	40·75	38·78	0·48	36·20	32·87	28·48	22·42	12·34
0·49	42·57	41·97	40·95	39·46	37·46	0·49	34·83	31·41	26·87	20·46	8·60

TABLE 2.2 (continued) 13

d/b	E → 5% H/b → 0.475	10% 0.450	15% 0.425	20% 0.400	25% 0.375	d/b	E → 30% H/b → 0.350	35% 0.325	40% 0.300	45% 0.275
0.50	41.36	40.75	39.71	38.20	36.16	0.50	33.48	29.96	25.25	18.41
0.51	40.17	39.55	38.50	36.96	34.88	0.51	32.14	28.52	23.61	16.23
0.52	39.00	38.37	37.30	35.75	33.62	0.52	30.81	27.08	21.94	13.86
0.53	37.85	37.22	36.13	34.55	32.38	0.53	29.50	25.64	20.23	11.15
0.54	36.73	36.08	34.98	33.36	31.15	0.54	28.19	24.19	18.46	7.78
0.55	35.63	34.97	33.85	32.20	29.94	0.55	26.89	22.73	16.61	
0.56	34.54	33.86	32.73	31.05	28.73	0.56	25.59	21.25	14.65	
0.57	33.48	32.80	31.63	29.92	27.54	0.57	24.30	19.75	12.51	
0.58	32.43	31.74	30.55	28.80	26.35	0.58	23.00	18.21	10.07	
0.59	31.40	30.70	29.48	27.69	25.18	0.59	21.69	16.61	7.02	
0.60	30.39	29.67	28.43	26.59	24.01	0.60	20.38	14.95		
0.61	29.39	28.66	27.39	25.51	22.84	0.61	19.05	13.18		
0.62	28.41	27.67	26.37	24.43	21.67	0.62	17.69	11.25		
0.63	27.45	26.69	25.36	23.36	20.50	0.63	16.31	9.05		
0.64	26.50	25.72	24.36	22.30	19.33	0.64	14.87	6.31		
0.65	25.57	24.77	23.37	21.25	18.15	0.65	13.38			
0.66	24.64	23.82	22.39	20.19	16.95	0.66	11.79			
0.67	23.74	22.89	21.41	19.14	15.73	0.67	10.06			
0.68	22.84	21.98	20.45	18.09	14.49	0.68	8.09			
0.69	21.96	21.07	19.49	17.04	13.20	0.69	5.63			
0.70	21.09	20.17	18.54	15.98	11.86					
0.71	20.23	19.28	17.60	14.90	10.44					
0.72	19.38	18.41	16.65	13.81	8.90					
0.73	18.54	17.54	15.71	12.70	7.15					
0.74	17.72	16.67	14.76	11.56	4.97					
0.75	16.90	15.82	13.81	10.37						
0.76	16.09	14.96	12.86	9.11						
0.77	15.30	14.12	11.89	7.75						
0.78	14.51	13.27	10.91	6.22						
0.79	13.73	12.43	9.90	4.32						
0.80	12.95	11.59	8.86							
0.81	12.19	10.75	7.76							
0.82	11.43	9.90	6.58							
0.83	10.68	9.04	5.26							
0.84	9.93	8.16	3.64							
0.85	9.19	7.27								
0.86	8.45	6.33								
0.87	7.71	5.34								
0.88	6.97	4.24								
0.89	6.22	2.91								
0.90	5.47									
0.91	4.70									
0.92	3.90									
0.93	3.04									
0.94	2.05									

d is the diameter of centre conductor, b is the inner diameter of outer conductor, $H = \frac{1}{2}(b-d)$, Z_0 is the characteristic impedance. κ is the dielectric constant of the medium filling the line interior, and E is the eccentricity $= 100(1-2H/b)$ per cent.

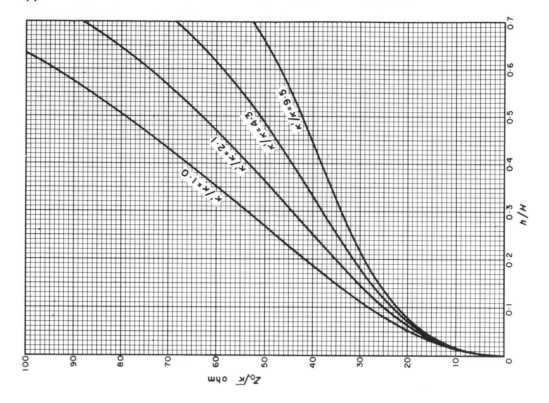

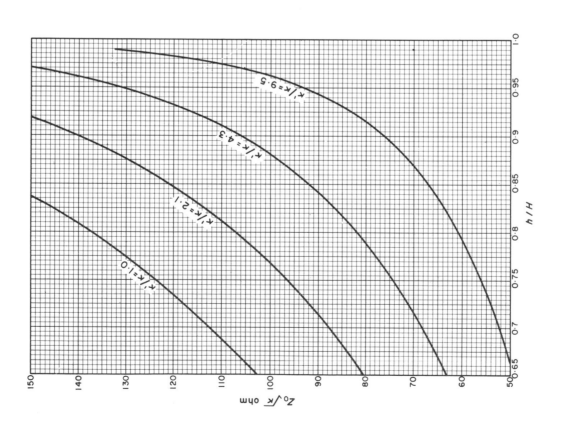

Fig. 2.7(a) and (b). Characteristic impedance of "round microstrip", with substrate dielectric constant as parameter

$$S = \sum_{n=0}^{\infty} (-D)^{n+1} \ln\left(1 + \frac{2}{n+x}\right) \quad (2.4.12)$$

and

$$D = \frac{1 - (\kappa/\kappa')}{1 + (\kappa/\kappa')} \quad (2.4.13)$$

These formulae were used to calculate the data graphed on Fig. 2.7 and tabulated in Table 2.3. Three commonly-used values of κ' have been assumed: $\kappa' = 2\cdot1$, corresponding to P.T.F.E.; $\kappa' = 4\cdot3$, corresponding to certain fibreglass laminates used in printed circuit boards; and $\kappa' = 9\cdot5$, corresponding to a typical alumina substrate widely used for microwave microstrip structures.

2.5 THE UNSCREENED TWIN-WIRE LINE

This, the second of the "interlopers" mentioned in Section 2.1, is included purely because of its relationship to the lines previously described, and to be described later. Because of its completely unscreened nature, microwave energy is immediately lost by radiation, and virtually no transmission can be obtained at frequencies in the microwave range.

Physically, the line consists of two parallel cylindrical wires, each of diameter d, with their axes separated by a distance $2H$, embedded within an (infinite) insulating medium of dielectric constant κ, as shown in cross-section in Fig. 2.8. It

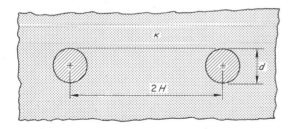

Fig. 2.8. Cross-section of unscreened twin-wire line, embedded in an infinite medium of dielectric constant κ

is immediately apparent that the structure can be regarded as two of the "single-wire-above-ground" lines (Section 2.4) with a common ground-plane located symmetrically between the two wires. Since this ground-plane constitutes an electric wall, its presence does not disturb or affect the field that already exists in its absence.

Since the two-wire system can easily be analysed

by elementary electrostatic techniques, its characteristic impedance can quickly be obtained from its electrostatic capacitance. In many texts, the characteristic impedance obtained in this way is used to derive that for the single-wire-above-ground, simply by use of a factor 1/2. Here, we reverse the process, and by doubling the result already obtained above (equation 2.4.2(a)), the characteristic impedance of the "unscreened-twin" is given by (see also [1], [7])

$$Z_0\sqrt{\kappa} = 119\cdot904 \ln\left\{\frac{2H}{d} + \sqrt{\frac{4H^2}{d^2} - 1}\right\} \text{ ohm.} \quad (2.5.1)$$

No tabulation or graph is given for equation (2.5.1), since this line is of such little practical use at microwave frequencies. If such data are required, they can quickly be obtained by doubling the values given in Table 2.3 or Fig. 2.7.

TABLE 2.3 CHARACTERISTIC IMPEDANCE OF "ROUND MICROSTRIP" AS A FUNCTION OF THE LINE PARAMETERS

			$\kappa'/\kappa \rightarrow 1\cdot0$	$2\cdot1$	$4\cdot3$	$9\cdot5$
h/H	d/H	d/h	$Z_0\sqrt{\kappa}$	$Z_0\sqrt{\kappa}$	$Z_0\sqrt{\kappa}$	$Z_0\sqrt{\kappa}$
0·005	1·99	398·00	6·01	5·85	5·77	5·73
0·015	1·97	131·33	10·45	9·97	9·74	9·61
0·025	1·95	78·00	13·55	12·75	12·35	12·14
0·035	1·93	55·14	16·10	14·98	14·41	14·10
0·045	1·91	42·44	18·33	16·89	16·15	15·75
0·055	1·89	34·36	20·36	18·59	17·68	17·17
0·065	1·87	28·77	22·23	20·14	19·05	18·44
0·075	1·85	24·67	23·98	21·58	20·30	19·58
0·085	1·83	21·53	25·65	22·93	21·46	20·62
0·095	1·81	19·05	27·24	24·20	22·54	21·59
0·105	1·79	17·05	28·76	25·41	23·55	22·49
0·115	1·77	15·39	30·24	26·57	24·52	23·33
0·125	1·75	14·00	31·68	27·69	25·44	24·12
0·135	1·73	12·81	33·07	28·77	26·32	24·87
0·145	1·71	11·79	34·44	29·82	27·16	25·58
0·155	1·69	10·90	35·78	30·84	27·98	26·26
0·165	1·67	10·12	37·09	31·84	28·76	26·91
0·175	1·65	9·43	38·39	32·83	29·53	27·53
0·185	1·63	8·81	39·67	33·79	30·28	28·13
0·195	1·61	8·26	40·93	34·74	31·01	28·71

TABLE 2.3 (continued)

h/H	d/H	d/h	$\kappa'/\kappa \to 1\cdot0$ $Z_0\sqrt{\kappa}$	$2\cdot1$ $Z_0\sqrt{\kappa}$	$4\cdot3$ $Z_0\sqrt{\kappa}$	$9\cdot5$ $Z_0\sqrt{\kappa}$	h/H	d/H	d/h	$\kappa'/\kappa \to 1\cdot0$ $Z_0\sqrt{\kappa}$	$2\cdot1$ $Z_0\sqrt{\kappa}$	$4\cdot3$ $Z_0\sqrt{\kappa}$	$9\cdot5$ $Z_0\sqrt{\kappa}$
0·205	1·59	7·76	42·18	35·67	31·72	29·27	0·655	0·69	1·05	103·49	81·10	63·50	49·57
0·215	1·57	7·30	43·42	36·59	32·42	29·81	0·665	0·67	1·01	105·36	82·51	64·49	50·15
0·225	1·55	6·89	44·64	37·51	33·10	30·34	0·675	0·65	0·96	107·29	83·97	65·51	50·75
0·235	1·53	6·51	45·87	38·41	33·78	30·85	0·685	0·63	0·92	109·27	85·47	66·56	51·36
0·245	1·51	6·16	47·08	39·31	34·44	31·35	0·695	0·61	0·88	111·30	87·02	67·64	52·00
0·255	1·49	5·84	48·29	40·21	35·10	31·84	0·705	0·59	0·84	113·39	88·62	68·76	52·67
0·265	1·47	5·55	49·49	41·10	35·75	32·31	0·715	0·57	0·80	115·55	90·27	69·92	53·36
0·275	1·45	5·27	50·69	41·98	36·39	32·78	0·725	0·55	0·76	117·79	91·97	71·12	54·08
0·285	1·43	5·02	51·89	42·86	37·03	33·23	0·735	0·53	0·72	120·09	93·74	72·37	54·83
0·295	1·41	4·78	53·09	43·75	37·66	33·68	0·745	0·51	0·68	122·48	95·57	73·66	55·61
0·305	1·39	4·56	54·29	44·63	38·29	34·12	0·755	0·49	0·65	124·96	97·47	75·02	56·42
0·315	1·37	4·35	55·49	45·51	38·92	34·56	0·765	0·47	0·61	127·53	99·45	76·43	57·28
0·325	1·35	4·15	56·69	46·39	39·55	34·99	0·775	0·45	0·58	130·21	101·52	77·90	58·18
0·335	1·33	3·97	57·90	47·27	40·17	35·41	0·785	0·43	0·55	133·00	103·67	79·44	59·13
0·345	1·31	3·80	59·11	48·16	40·80	35·83	0·795	0·41	0·52	135·92	105·93	81·07	60·13
0·355	1·29	3·63	60·32	49·05	41·42	36·24	0·805	0·39	0·48	138·98	108·30	82·77	61·19
0·365	1·27	3·48	61·54	49·94	42·05	36·66	0·815	0·37	0·45	142·20	110·80	84·57	62·32
0·375	1·25	3·33	62·77	50·84	42·67	37·06	0·825	0·35	0·42	145·59	113·43	86·48	63·51
0·385	1·23	3·19	64·00	51·74	43·30	37·47	0·835	0·33	0·39	149·17	116·22	88·51	64·79
0·395	1·21	3·06	65·24	52·65	43·93	37·88	0·845	0·31	0·37	152·96	119·19	90·66	66·17
0·405	1·19	2·94	66·49	53·57	44·57	38·28	0·855	0·29	0·34	157·01	122·35	92·97	67·64
0·415	1·17	2·82	67·75	54·49	45·20	38·68	0·865	0·27	0·31	161·33	125·74	95·46	69·24
0·425	1·15	2·71	69·02	55·43	45·85	39·08	0·875	0·25	0·29	165·99	129·39	98·14	70·98
0·435	1·13	2·60	70·30	56·37	46·49	39·49	0·885	0·23	0·26	171·02	133·35	101·06	72·89
0·445	1·11	2·49	71·59	57·32	47·15	39·89	0·895	0·21	0·23	176·51	137·67	104·25	74·99
0·455	1·09	2·40	72·89	58·28	47·81	40·30	0·905	0·19	0·21	182·54	142·43	107·78	77·33
0·465	1·07	2·30	74·21	59·25	48·47	40·70	0·915	0·17	0·19	189·23	147·72	111·72	79·95
0·475	1·05	2·21	75·55	60·23	49·15	41·11	0·925	0·15	0·16	196·76	153·69	116·17	82·94
0·485	1·03	2·12	76·90	61·23	49·83	41·52	0·935	0·13	0·14	205·36	156·51	121·27	86·39
0·495	1·01	2·04	78·26	62·24	50·52	41·94	0·945	0·11	0·12	215·40	168·48	127·26	90·47
0·505	0·99	1·96	79·65	63·26	51·22	42·36	0·955	0·09	0·09	227·44	178·08	134·49	95·42
0·515	0·97	1·88	81·05	64·30	51·94	42·78	0·965	0·07	0·07	242·52	190·10	143·57	101·70
0·525	0·95	1·81	82·48	65·36	52·66	43·21	0·975	0·05	0·05	262·70	206·23	155·80	110·21
0·535	0·93	1·74	83·92	66·43	53·39	43·65	0·985	0·03	0·03	293·33	230·75	174·45	123·28
0·545	0·91	1·67	85·39	67·52	54·14	44·09							
0·555	0·89	1·60	86·88	68·63	54·90	44·53							
0·565	0·87	1·54	88·40	69·76	55·68	44·99							
0·575	0·85	1·48	89·94	70·92	56·47	45·45							
0·585	0·83	1·42	91·52	72·09	57·28	45·93							
0·595	0·81	1·36	93·12	73·29	58·11	46·41							
0·605	0·79	1·31	94·75	74·52	58·95	46·90							
0·615	0·77	1·25	96·42	75·77	59·82	47·41							
0·625	0·75	1·20	98·13	77·05	60·70	47·93							
0·635	0·73	1·15	99·87	78·37	61·61	48·46							
0·645	0·71	1·10	101·66	79·71	62·54	49·00							

Z_0 is the characteristic impedance, h is the height of dielectric substrate, d is the diameter of line conductor, H is the $h+d/2$, κ' is the dielectric constant of substrate, and κ is the dielectric constant of surrounding medium.

REFERENCES

1. Jackson, W. *High-Frequency Transmission Lines.*
   pp. 45–96, 3rd edn, Methuen & Co. Ltd., Lond. (1958).
2. Moon, P. and Spencer, D. E. *Foundations of Electrodynamics.*
   p. 192, Van Nostrand, New Jersey (1960).
3. Poincelot, P. "Théorie du cable coaxial", *Annls Télécommun.*
   17, Nos. 5–6, 94–98 (1962).
4. Poincelot, P. "Théorie de la ligne coaxiale", *Câbles Trans.*
   17, No. 4, 227–238 (1963).
5. Moreno, T. *Microwave Transmission Design Data.*
   Dover, New York (1958).
6. King, R. W. P. *Transmission-line Theory.*
   pp. 31–33, McGraw-Hill, New York (1955).
7. *Reference Data for Radio Engineers.*
   5th edn, Section 22–24, Howard W. Sams & Co., New York (1968).
8. *Microwave J.: Microwave Engineers' Technical and Buyers' Guide Edition.*
   p. 34, Horizon House, Dedham, Mass. (Feb. 1969).
9. Hasse, J. A. "Eccentric-Line Impedance Nomograph". *Electronics.*
   p. 190 (Sept. 1956).
10. Abaka, E. and Baier, W. "TE and TM modes in transmission lines with circular outer conductor and eccentric circular inner conductor". *Electronic Lett.*
    5, pp. 251–252, 11 (May 29th 1969).
11. Kaden, H. "Leitungs-und Kopplungskonstanten bei Streifenleitungen". *AEÜ.*
    21, pp. 109–111 (1967).
12. James, G. and James, R. C. *Mathematics Dictionary.*
    Formula 400 on page 466, Van Nostrand, Princeton, N.J. (1959).
13. Igushkin, L. P. and Urazakov, E. I. "Excitation of the TEM Wave in an imperfectly conducting Coaxial Line", *Radio Engng electron. Phys.*
    14, pp. 776–778, 5 (1969).
14. Yee, H. Y. and Audeh, N. F. "Cutoff Frequencies of Eccentric Waveguides", *Trans. I.E.E.E.*
    MTT-14, pp. 487–493 (Oct. 1966).

SUMMARY OF USEFUL FORMULAE

Structure	Cross-section	Impedance formulae	Text equation no.
Coaxial Line	 **Fig. 2.1**	$Z_0\sqrt{\kappa} = 59{\cdot}952 \ln (b/d)$	(2.2.2)
Eccentric Coaxial Line	 **Fig. 2.3**	$Z_0\sqrt{\kappa} = 59{\cdot}952 \ln (X+\sqrt{X^2-1})$ $X = \dfrac{1}{2}\left[\dfrac{d}{b}+\dfrac{4H}{d}\left(1-\dfrac{H}{b}\right)\right]$	(2.3.1) (2.3.2)
Single-wire-above-ground	 **Fig. 2.5**	$Z_0\sqrt{\kappa} = 59{\cdot}952 \ln (X+\sqrt{X^2-1})$ $X = 2H/d$	(2.4.2a) (2.4.1)
"Round Microstrip"	 **Fig. 2.6**	$Z_0\sqrt{\kappa} \doteqdot 59{\cdot}952\sqrt{L(L+S)}$ $L = \ln\left[1+\dfrac{1}{2x}(1+\sqrt{1+4x})\right]$ $x = d/4h$ $S = \displaystyle\sum_{n=0}^{\infty} (-D)^{n+1} \ln\left(1+\dfrac{2}{n+x}\right)$ $D = \dfrac{\kappa'-\kappa}{\kappa'\;\kappa}$	(2.4.9) (2.4.10) (2.4.11) (2.4.12) (2.4.13)
Unscreened twin-wire line	 **Fig. 2.8**	$Z_0\sqrt{\kappa} = 119{\cdot}904 \ln (X+\sqrt{X^2-1})$ $X = 2H/d$	(2.5.1)

3

TRANSMISSION LINES UTILIZING CONDUCTORS OF RECTANGULAR CROSS-SECTION

3.1 INTRODUCTION

Unlike some of the lines discussed in Chapter 2, all of those to be treated in this chapter can unquestionably be described as *microwave* transmission lines.

Some, of course, are more widely used than others: particular mention should be made of the widespread use of triplate (Section 3.5) striplines in the production of cheap, high-quality, miniaturized microwave systems, particularly for use in the low GHz range of frequencies. Microstrip (Section 3.6), too, has recently been under intense investigation and development as a means of producing microwave integrated circuits for use at frequencies as high, in some instances, as 70–100 GHz.

Each of the lines described below has its own set of advantages and disadvantages, which dictate its suitability for one specialized application or another—the purpose here is not to discuss these, but to present the performance characteristics of each type so that the systems designer will be enabled to make his own choice.

Following the pattern already laid down, the discussion opens with a treatment of the most symmetric form, namely, the square coaxial line.

However, the analysis of this, and of some of the related types of line, involves the use of elliptic integrals, so, for future reference, the chapter begins with a brief discussion of these functions and a statement of relevant formulae, together with tabulations of useful numerical data.

3.2 ELLIPTIC INTEGRALS IN TRANSMISSION-LINE ANALYSIS

The analysis, by use of conformal transformation techniques [1, 2], of transmission lines whose cross-sections include right-angled configurations involves, at one stage or another, the use of elliptic integrals. Since a complete treatment of these is obviously inappropriate for inclusion in the present text, the discussion will be limited to the most frequently encountered example (the interested reader will find a detailed and comprehensive description of elliptic integrals in [3]).

This is the "complete elliptic integral of the 1st kind", usually denoted by $K(k)$, which is defined as follows:

$$K(k) = \int_0^{\pi/2} \frac{d\phi}{\sqrt{1 - k^2 \sin^2 \phi}} \tag{3.2.1}$$

$$= \int_0^1 \frac{dx}{\sqrt{(1 - x^2)(1 - k^2 x^2)}} \tag{3.2.2}$$

k is known as the "modulus" of K.

The associated "complementary" function $K'(k)$ is defined by

$$K'(k) = K(k') \qquad (3.2.3)$$

where

$$k' = (1-k^2)^{1/2} \qquad (3.2.4)$$

The evaluation of K and K' involves extremely tedious hand-calculation and/or computer programming, and in general it is preferable to make use of available tables, e.g. [4]. However, in transmission-line applications, these elliptic functions are most frequently encountered in the form of the ratio K/K', and there are in existence some simple, and very accurate, closed-form formulae for this ratio in terms of k. Which formula is to be used depends upon the accuracy required: Hilberg [5] has quoted three versions, as follows:

(a) For $1 \leqslant K/K' \leqslant \infty$ and $0.5 \leqslant k^2 \leqslant 1$

$$\frac{K}{K'} \doteqdot \frac{2}{\pi} \ln\left(2\sqrt{\frac{1+k}{1-k}}\right) = F_1(k), \text{ say} \qquad (3.2.5)$$

For $0 \leqslant K/K' \leqslant 1$ and $0 \leqslant k^2 \leqslant 0.5$

$$\frac{K}{K'} \doteqdot \frac{1}{F_1(k')} \qquad (3.2.6)$$

Claimed accuracy is better than three parts per thousand.

(b) The following approximation appears to have been first derived by Whittaker and Watson [6]: Hilberg [5] gives an independent derivation.

For $1 \leqslant K/K' \leqslant \infty$ and $0.5 \leqslant k^2 \leqslant 1$

$$\frac{K}{K'} \doteqdot \frac{1}{\pi} \ln 2\left(\frac{1+\sqrt{k}}{1-\sqrt{k}}\right), = F_2(k) \qquad (3.2.7)$$

For $0 \leqslant K/K' \leqslant 1$ and $0 \leqslant k^2 \leqslant 0.5$

$$\frac{K}{K'} \doteqdot \frac{1}{F_2(k')} \qquad (3.2.8)$$

Here, Hilberg claims an accuracy of three parts per million, whereas Whittaker and Watson quote eight parts per million. However, for all practical usage, this difference in accuracy is obviously of purely academic importance!

(c) The following approximation is also derived by Hilberg.

For $1 \leqslant K/K' \leqslant \infty$ and $0.5 \leqslant k^2 \leqslant 1$

For $0 \leqslant K/K' \leqslant 1$ and $0 \leqslant k^2 \leqslant 0.5$

$$\frac{K}{K'} \doteqdot \frac{1}{F_3(k')} \qquad (3.2.10)$$

The accuracy claimed here is four parts in 10^{12}, which is negligibly different from exactitude, and is in fact higher than in any available tables, according to Hilberg.

For the reader's convenience, values of K/K' versus k are tabulated in Table 3.1, as calculated from equations (3.2.9) and (3.2.10). Note that in each of the above pairs of expressions, one is obtained from the other by taking its reciprocal, after substituting k' for k or vice versa.

For general use, it would appear that the expressions given in paragraph (b) above offer the best combination of high accuracy with algebraic simplicity: in particular they can readily be inverted to yield k or k' in terms of K/K', e.g. from equation (3.2.7) there results

$$k = \left(\frac{e^x - 2}{e^x + 2}\right)^2 \qquad (3.2.11)$$

where

$$x = \pi K/K' \qquad (3.2.12)$$

and similarly, from equation (3.2.8),

$$k' = \left(\frac{e^x - 2}{e^x + 2}\right)^2 \qquad (3.2.13)$$

where now

$$x = \pi K'/K \qquad (3.2.14)$$

Having thus established the basic mathematical relationships needed, the main topics of this chapter can now be developed.

3.3   THE SQUARE COAXIAL LINE

From the cross-sectional view shown in Fig. 3.1, this line is obviously directly analogous to the cylindrical coaxial line treated in Section 2.2.

Because of its uniform symmetry, the square line is much simpler to analyse than the general rectangular line, by the use of conformal transformation techniques [1, 2].

The transformations appropriate to the present

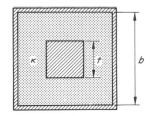

Fig. 3.1. Cross-section of the square coaxial line. κ is the dielectric constant of the medium filling the line exterior

configuration appear to have been first derived by Bowman [7], who later ([3], pp. 99–104) obtained an expression for the capacitance per unit length of the square coaxial line:

$$C = 8\kappa\varepsilon_0 \frac{K(k)}{K'(k)} \text{ farads/metre.} \qquad (3.3.1)$$

Using this to derive the formula for characteristic impedance, there results

$$Z_0\sqrt{\kappa} = 47{\cdot}086 \frac{K'(k)}{K(k)} \text{ ohm} \qquad (3.3.2)$$

where κ is the dielectric constant of the medium filling the line.

The modulus k is related to the line dimensions via the following equations:

$$k = \frac{(p-q)^2}{(p+q)^2} \qquad (3.3.3)$$

$$p^2 + q^2 = 1 \qquad (3.3.4)$$

where

$$\frac{K(p)}{K'(p)} = \frac{1-t/b}{1+t/b}. \qquad (3.3.5)$$

This result is also quoted by Conning [8] and Green [9].

A similar, and somewhat simpler, derivation has been given by Anderson [10] (see Section 3.4.3), which is, however, less convenient for computational purposes.

Results calculated from equations (3.3.1)–(3.3.5) using equation (3.2.9) to evaluate $K'(k)/K(k)$, are listed in the second and fifth columns of Table 3.2: these "exact" values (within the accuracy of the elliptic function ratio) may be compared with those tabulated by Green [11], which were derived from a numerical finite-difference analysis of the structure.

Cockcroft [21] has given detailed analyses of several right-angled electrode configurations, and

his equation (37) can be used to derive an algebraically simple expression for the characteristic impedance of square coaxial line, namely (cf. [9])

$$Z_0\sqrt{\kappa} = \frac{47{\cdot}086\,(1-t/b)}{(0{\cdot}279+0{\cdot}721\,t/b)} \text{ ohm.} \qquad (3.3.6)$$

This rather simple expression yields values of Z_0 which, for $t/b > 0{\cdot}25$, are within 1 per cent of those calculated from Bowman's exact formulae (see Table 3.2). In the many engineering applications where such an accuracy is acceptable, formula (3.3.6) is obviously much easier to use than formula (3.3.2), even when using the approximations for K/K' given in Section 3.2; with the added advantage that it can readily be inverted to yield a simple expression for t/b in terms of Z_0.

Results calculated from both formulae are tabulated in Table 3.2, to facilitate comparison, and for use in design applications. A graphical display of Bowman's results is given in Fig. 3.2.

An even simpler formula, empirically derived from the results of his finite-difference analysis [11], is quoted by Green, claiming an accuracy of within 0·5 per cent for $t/b \leqslant 0{\cdot}5$, namely

$$Z_0\sqrt{\kappa} = 136{\cdot}7 \log_{10}(0{\cdot}9259\,b/t) \text{ ohm.} \qquad (3.3.7)$$

Conclusion to Section 3.3

For accurate design work use the data listed in Table 3.2, or use formula (3.3.2) in conjunction with the elliptic integral formulae of Section 3.2.

For approximate calculations, the simple formulae (3.3.6) or (3.3.7) may be used.

3.4 THE RECTANGULAR COAXIAL LINE

This is the "basic form" from which almost all of the other lines considered in this chapter can be derived, by suitable transformations and/or limiting processes which will be indicated where appropriate.

This basic structure consists of a conducting strip of rectangular cross-section (of width W and thickness t) which is (usually but not necessarily) axially and symmetrically located within an enclosing conducting shield, also of rectangular cross-section (of width W' and height b), as shown schematically in Fig. 3.3.

TABLE 3.1 VALUES OF THE COMPLETE ELLIPTIC INTEGRAL RATIO, K(k)/K'(k), AS A FUNCTION OF THE MODULUS k

k	K(k)/K'(k)	k	K(k)/K'(k)
0·005	0·238632	0·155	0·484133
0·010	0·262507	0·160	0·488978
0·015	0·281210	0·165	0·493772
0·020	0·296487	0·170	0·498517
0·025	0·309500	0·175	0·503217
0·030	0·321061	0·180	0·507874
0·035	0·331502	0·185	0·512492
0·040	0·341124	0·190	0·517072
0·045	0·350087	0·195	0·521618
0·050	0·358516	0·200	0·526130
0·055	0·366499	0·205	0·530612
0·060	0·374106	0·210	0·535064
0·065	0·381391	0·215	0·539490
0·070	0·388395	0·220	0·543890
0·075	0·395154	0·225	0·548266
0·080	0·401695	0·230	0·552620
0·085	0·408043	0·235	0·556953
0·090	0·414217	0·240	0·561267
0·095	0·420234	0·245	0·565562
0·100	0·426109	0·250	0·569841
0·105	0·431855	0·255	0·574104
0·110	0·437483	0·260	0·578353
0·115	0·443003	0·265	0·582588
0·120	0·448423	0·270	0·586810
0·125	0·453751	0·275	0·591022
0·130	0·458994	0·280	0·595223
0·135	0·464158	0·285	0·599415
0·140	0·469249	0·290	0·603598
0·145	0·474272	0·295	0·607774
0·150	0·479232	0·300	0·611943
0·305	0·616107	0·505	0·786216
0·310	0·620266	0·510	0·790754
0·315	0·624420	0·515	0·795316
0·320	0·628571	0·520	0·799903
0·325	0·632720	0·525	0·804516
0·330	0·636867	0·530	0·809155
0·335	0·641013	0·535	0·813822
0·340	0·645158	0·540	0·818517
0·345	0·649304	0·545	0·823241
0·350	0·653451	0·550	0·827996

k	K(k)/K'(k)	k	K(k)/K'(k)
0·355	0·657600	0·555	0·832782
0·360	0·661751	0·560	0·837600
0·365	0·665906	0·565	0·842451
0·370	0·670064	0·570	0·847337
0·375	0·674227	0·575	0·852259
0·380	0·678395	0·580	0·857217
0·385	0·682569	0·585	0·862212
0·390	0·686749	0·590	0·867247
0·395	0·690937	0·595	0·872322
0·400	0·695132	0·600	0·877438
0·405	0·699336	0·605	0·882597
0·410	0·703549	0·610	0·887800
0·415	0·707772	0·615	0·893048
0·420	0·712005	0·620	0·898344
0·425	0·716249	0·625	0·903688
0·430	0·720505	0·630	0·909083
0·435	0·724773	0·635	0·914529
0·440	0·729055	0·640	0·920028
0·445	0·733350	0·645	0·925582
0·450	0·737659	0·650	0·931194
0·455	0·741984	0·655	0·936864
0·460	0·746324	0·660	0·942595
0·465	0·750681	0·665	0·948389
0·470	0·755055	0·670	0·954249
0·475	0·759447	0·675	0·960175
0·480	0·763858	0·680	0·966171
0·485	0·768288	0·685	0·972239
0·490	0·772738	0·690	0·978382
0·495	0·777209	0·695	0·984602
0·500	0·781701	0·700	0·990902
0·705	0·997285	0·905	1·395481
0·710	1·003754	0·910	1·413547
0·715	1·010313	0·915	1·432593
0·720	1·016964	0·920	1·452739
0·725	1·023712	0·925	1·474127
0·730	1·030560	0·930	1·496930
0·735	1·037512	0·935	1·521358
0·740	1·044572	0·940	1·547672
0·745	1·051745	0·945	1·576200
0·750	1·059036	0·950	1·607367
0·755	1·066449	0·955	1·641729
0·760	1·073990	0·960	1·680043
0·765	1·081665	0·965	1·723366
0·770	1·089479	0·970	1·773250
0·775	1·097439	0·975	1·832097

TABLE 3.1 (continued)

k	K(k)/K′(k)	k	K(k)/K′(k)
0·780	1·105552	0·980	1·903935
0·785	1·113824	0·985	1·996314
0·790	1·122264	0·990	2·126181
0·795	1·130881	0·995	2·347617
0·800	1·139682		
0·805	1·148678	0·9900	2·126181
0·810	1·157879	0·9905	2·142588
0·815	1·167297	0·9910	2·159878
0·820	1·176943	0·9915	2·178152
0·825	1·186830	0·9920	2·197530
0·830	1·196973	0·9925	2·218153
0·835	1·207388	0·9930	2·240194
0·840	1·218091	0·9935	2·263863
0·845	1·229101	0·9940	2·289422
0·850	1·240438	0·9945	2·317198
0·855	1·252126	0·9950	2·347617
0·860	1·264188	0·9955	2·381234
0·865	1·276653	0·9960	2·418805
0·870	1·289551	0·9965	2·461388
0·875	1·302916	0·9970	2·510538
0·880	1·316787	0·9975	2·568648
0·885	1·331208	0·9980	2·639760
0·890	1·346227	0·9985	2·731412
0·895	1·361902	0·9990	2·860549
0·900	1·378295		

3.4.1   Analysis of the General Rectangular Line

Although the structure is of a pleasing and simple symmetry, an exact general analysis is extremely difficult, and until recently none had been given. However, in [12] Sato and Ikeda (in a contributed chapter) have given a fairly comprehensive coverage of several types of microwave transmission lines, and, using conformal transformation techniques, have derived exact formulae for the various line parameters. Among these one can find some rather complex, implicit, but exact, formulae which can be used to determine the exact characteristic impedance of rectangular coaxial line of arbitrary cross-sectional dimensions.

Using present notation, these are as follows:

$$\frac{W}{b} = \frac{I(\alpha_1, 1)}{I(0, \alpha_1) + I(1, 1/k)} \tag{3.4.1}$$

$$\frac{t}{b} = \frac{I(0, \alpha_1)}{I(0, \alpha_1) + I(1, 1/k)} \tag{3.4.2}$$

$$\frac{W'}{b} = \frac{I(1/k, 1/\alpha_2)}{I(0, \alpha_1) + I(1, 1/k)} \tag{3.4.3}$$

where

$$I(t_1, t_2) = \int_{t_1}^{t_2} \sqrt{\left| \frac{(t^2 - \alpha_1^2)}{(t^2 - 1)(t^2 - 1/k^2)(t^2 - 1/\alpha_2^2)} \right|} \, dt.$$

In a masterpiece of understatement, the authors baldly state that "These relations can be used to determine k." It is indeed true that, once k is known, the characteristic impedance is obtained from the following relatively simple formula,

$$Z_0 \sqrt{\kappa} = 29 \cdot 976\pi \frac{K'(k)}{K(k)} \text{ ohm} \tag{3.4.4}$$

but the elliptic-type functions involved in determining k from equations (3.4.1)–(3.4.3) are "non-standard" and untabulated. It is therefore unfortunately the case that although an "exact solution" is available, as indicated above, the labour involved in deriving therefrom numerical results of practical use is at least equal to, and probably greater than, that involved in obtaining similar results by purely numerical techniques, such as relaxation methods. Such methods have been adopted by several authors, and some of the most useful results obtained are given later. The "exact" approach, based upon the use of conformal transformations, has also been used by Lin and Chung [13] in an extensive analysis of several transmission lines, including the rectangular coaxial line: but all their results are in terms of upper and lower bounds on the characteristic impedance, and since the formulae obtained are still rather complex functions of elliptic functions, they are not of great practical utility.

Bräckelmann [14] has used a semi-analytic, semi-numerical approach to arrive at a series representation of the characteristic impedance of rectangular coaxial line, which is extremely general in that the cross-sectional dimensions can be completely arbitrary, and further, the axis of the strip conductor need not coincide with the axis of the rectangular shield. Such an analysis is obviously of great value in assessing the effects of dimensional tolerances. Reference [14] contains a comprehensive set of graphs of Z_0 as a

TABLE 3.2 THE CHARACTERISTIC IMPEDANCE OF SQUARE COAXIAL LINE AS A FUNCTION OF THE LINE PARAMETERS. Z_0 = CHARACTERISTIC IMPEDANCE (OHM). THE "BOWMAN" VALUES ARE EXACT; THE "COCKCROFT" VALUES ARE CALCULATED FROM A SIMPLE APPROXIMATION FORMULA (SEE TEXT). FOR DEFINITION OF PARAMETERS, SEE FIG. 3.1

t/b	Bowman $Z_0\sqrt{\kappa}$	Cockcroft $Z_0\sqrt{\kappa}$	t/b	Bowman $Z_0\sqrt{\kappa}$	Cockcroft $Z_0\sqrt{\kappa}$
0·01	270·69	162·67	0·46	41·63	41·62
0·02	229·14	157·07	0·47	40·38	40·38
0·03	204·83	151·75	0·48	39·16	39·16
0·04	187·58	146·67	0·49	37·97	37·97
0·05	174·20	141·83	0·50	36·81	36·80
0·06	163·27	137·20	0·51	35·67	35·67
0·07	154·03	132·78	0·52	34·55	34·55
0·08	146·02	128·54	0·53	33·47	33·46
0·09	138·96	124·48	0·54	32·40	32·40
0·10	132·65	120·59	0·55	31·36	31·36
0·11	126·93	116·85	0·56	30·34	30·34
0·12	121·72	113·26	0·57	29·34	29·34
0·13	116·92	109·81	0·58	28·36	28·36
0·14	112·48	106·49	0·59	27·40	27·40
0·15	108·34	103·30	0·60	26·46	26·46
0·16	104·48	100·22	0·61	25·54	25·54
0·17	100·84	97·25	0·62	24·64	24·64
0·18	97·42	94·38	0·63	23·76	23·76
0·19	94·18	91·62	0·64	22·89	22·89
0·20	91·11	88·95	0·65	22·04	22·04
0·21	88·19	86·37	0·66	21·20	21·20
0·22	85·40	83·87	0·67	20·39	20·39
0·23	82·74	81·45	0·68	19·58	19·58
0·24	80·20	79·12	0·69	18·80	18·80
0·25	77·76	76·85	0·70	18·02	18·02
0·26	75·41	74·66	0·71	17·26	17·26
0·27	73·16	72·53	0·72	16·52	16·52
0·28	70·99	70·46	0·73	15·78	15·78
0·29	68·89	68·46	0·74	15·06	15·06
0·30	66·87	66·51	0·75	14·36	14·36
0·31	64·92	64·62	0·76	13·66	13·66
0·32	63·03	62·79	0·77	12·98	12·98
0·33	61·20	61·00	0·78	12·31	12·31
0·34	59·43	59·26	0·79	11·65	11·65
0·35	57·71	57·57	0·80	11·00	11·00

t/b	Bowman $Z_0\sqrt{\kappa}$	Cockcroft $Z_0\sqrt{\kappa}$	t/b	Bowman $Z_0\sqrt{\kappa}$	Cockcroft $Z_0\sqrt{\kappa}$
0·36	56·04	55·93	0·81	10·37	10·37
0·37	54·42	54·33	0·82	9·74	9·74
0·38	52·84	52·77	0·83	9·12	9·12
0·39	51·31	51·25	0·84	8·52	8·52
0·40	49·82	49·77	0·85	8·05	7·92
0·41	48·36	48·33	0·86	—	7·33
0·42	46·95	46·92	0·87	—	6·75
0·43	45·57	45·55	0·88	—	6·19
0·44	44·23	44·21	0·89	—	5·63
0·45	42·91	42·90	0·90	—	5·07

function of cross-sectional dimensions, but unfortunately these are on too small a scale for accurate reading. Bräckelmann also quotes a rather simple approximate formula for Z_0, namely

$$Z_0\sqrt{\kappa} = 59\cdot952\ \ln\left(\frac{1+W'/b}{W/b+t/b}\right) \text{ohm.} \qquad (3.4.5)$$

which is stated to be within 10 per cent for $t/b < 0\cdot3$ and $W/W' < 0\cdot8$.

Bräckelmann's general equations [14] can be programmed for computer evaluation, but the resulting programs are not an economic proposition (in terms of computer run times) for the preparation of tables or graphs, and are best used only where highly accurate data are specifically required. For most practical design purposes, the extensive graphs presented by Metcalf [20] (which are reproduced here by kind permission of the I.E.E.) will be found as accurate and useful as any: see Fig. 3.4(a)–(h).

As in the case of the coaxial line, it is necessary to ensure that the cross-sectional dimensions used in a particular application are such as will not allow the propagation of higher-order modes. Bräckelmann et al. [15] have evaluated the cut-off frequencies of many such modes, in rectangular coaxial lines whose cross-sectional dimensions correspond to TEM mode characteristic impedances of between 20 and 200 ohm, and further data are given by Baier [16, 17].

Many other authors have contributed numerical analyses of the general rectangular line, but an extensive discussion of these is inappropriate for inclusion in what is primarily intended to be a data handbook: the interested reader may consult [18–20] as being typical treatments, which also list numerous references for further study.

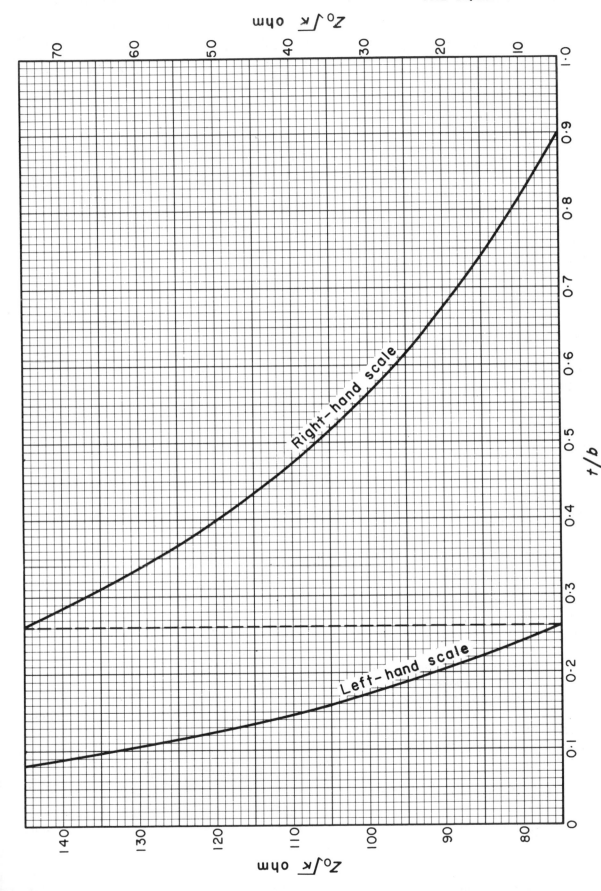

Fig. 3.2. The characteristic impedance of square coaxial line, as calculated from Bowman's exact formula. Note carefully the usage of the appropriate impedance scales. For definition of parameters, see Fig. 3.1

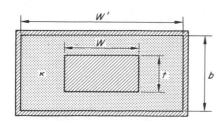

Fig. 3.3. Cross-section of the general rectangular coaxial line. κ is the dielectric constant of the medium filling the line interior

However, apart from exact or numerical analyses of the general case, it is also possible to perform exact analyses in certain special cases, depending upon the particular relationships between the various cross-sectional dimensions W, W', b, and t: an obvious example is the square coaxial line already discussed in Section 3.3, for which $W = W'$ and $t = b$.

In other instances, semi-empirical formulae can be derived, which yield results of varying degrees of accuracy.

Examples of both types are given in the paragraphs which follow.

3.4.2 The Equal-Gap Rectangular Line

Dimensional relationships: $W' - W = b - t$. (3.4.6)

In this special case, the spacing or gap between the conductors is the same in the vertical as in the horizontal plane.

As Green [11] has indicated, this makes it possible to derive an expression for the characteristic impedance by utilizing the results previously obtained for the square coaxial line (Section 3.3), as follows.

If the structure is divided into three sections by two imaginary vertical planes located at distances $t/2$ from the edges of the centre strip, it can be visualized that the two end-sections thus formed could be combined to yield a square coaxial structure. Hence, the total capacitance of the structure is, to a very good approximation, equal to that of the square coaxial line plus the simple parallel-plate capacitance of the centre section. Thus, using equation (3.3.1), the total capacitance is given by

$$C = 8\kappa\varepsilon_0 \frac{K(k)}{K'(k)} + 4\kappa\varepsilon_0 \frac{(W/b - t/b)}{1 - t/b} \text{ farads/metre} \quad (3.4.7)$$

whence the characteristic impedance is obtained as

$$Z_0\sqrt{\kappa} = \frac{47 \cdot 086}{\{K(k)/K'(k)\} + \frac{1}{2}\{(W/b - t/b)/(1 - t/b)\}} \text{ ohm} \quad (3.4.8)$$

where k is given in terms of t/b by equations (3.3.3) and (3.3.5).

The equal-gap case has also been treated by Chen [22], who uses Cockcroft's formulae [21] to derive an expression which can be written in the following form:

$$Z_0\sqrt{\kappa} = \frac{94 \cdot 172}{\{(W/b + t/b)/(1 - t/b)\} + 0 \cdot 558} \text{ ohm.} \quad (3.4.9)$$

Other cases analysed by Chen are discussed in Section 3.4.5.

3.4.3 Special case solved by Anderson

Anderson [10] has given an exact solution of the rectangular line, but under the rather restrictive conditions given below: it is included here mainly because of its value in checking the validity and accuracy of results obtained by the other, mainly numerical, techniques described in Section 3.4.1 and also because it yields the rather simpler expressions for the square coaxial line mentioned above.

Anderson's formulae for the rectangular line are as follows:

$$C = 2\kappa\varepsilon_0 \frac{K'(x)}{K(x)} \text{ farads/metre} \quad (3.4.10)$$

and hence

$$Z_0\sqrt{\kappa} = 188 \cdot 344 \frac{K(x)}{K'(x)} \text{ ohm} \quad (3.4.11)$$

The modulus x is obtained from the following formulae:

$$x = \frac{k_2}{k_1} \sqrt{\frac{(1 - k_1^2)}{(1 - k_2^2)}} \quad (3.4.12)$$

where

$$\frac{K(k_1)}{K'(k_1)} = \frac{W'/b + W/b}{1 - t/b} \quad (3.4.13)$$

and

$$\frac{K(k_2)}{K'(k_2)} = \frac{W'/b - W/b}{1 + t/b} \quad (3.4.14)$$

Further, equations (3.4.10) and (3.4.11) are valid *only* if the following condition is satisfied:

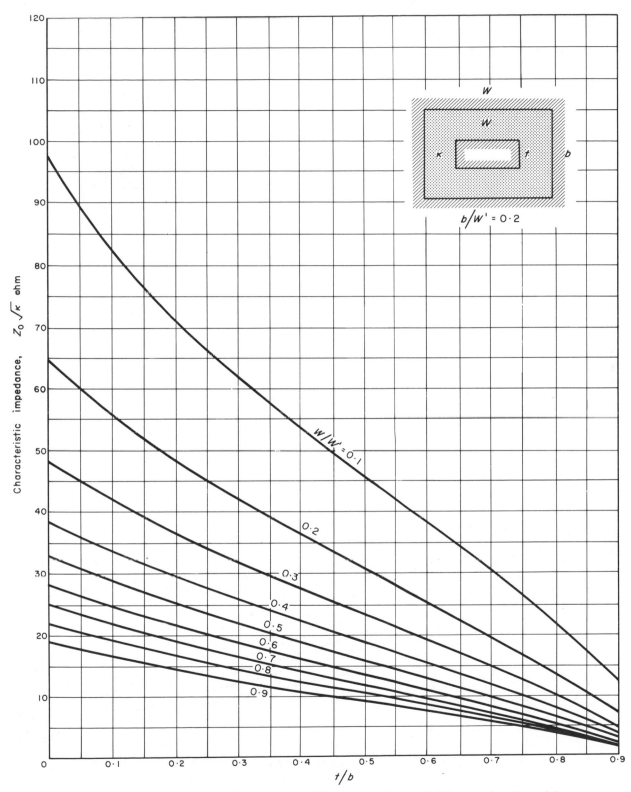

Fig. 3.4(a). The characteristic impedance of the rectangular coaxial line as a function of the cross-sectional dimensions, with $b/W' = 0.2$. (Courtesy of the I.E.E.)

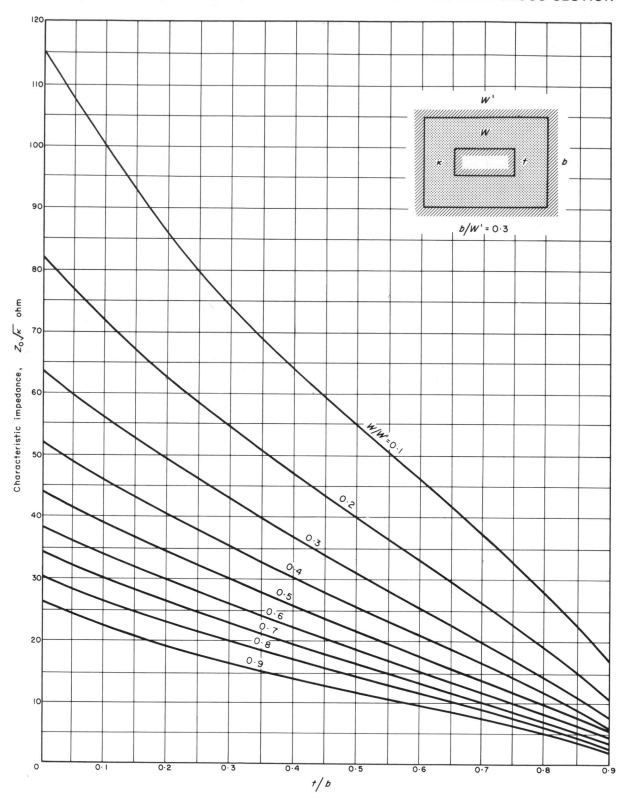

Fig. 3.4(b). The characteristic impedance of the rectangular coaxial line as a function of the cross-sectional dimensions, with $b/W' = 0.3$. (Courtesy of the I.E.E.)

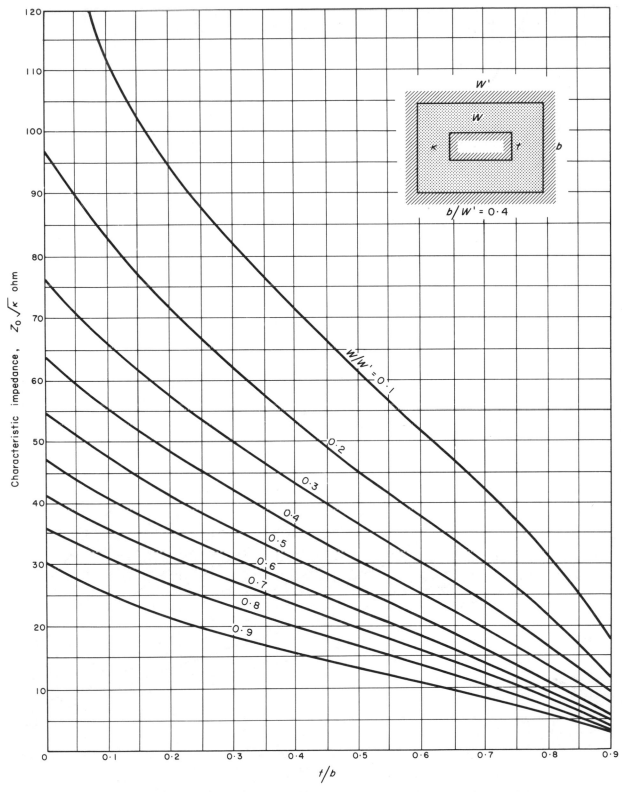

Fig. 3.4(c). The characteristic impedance of the rectangular coaxial line as a function of the cross-sectional dimensions, with $b/W' = 0.4$. **(Courtesy of the I.E.E.)**

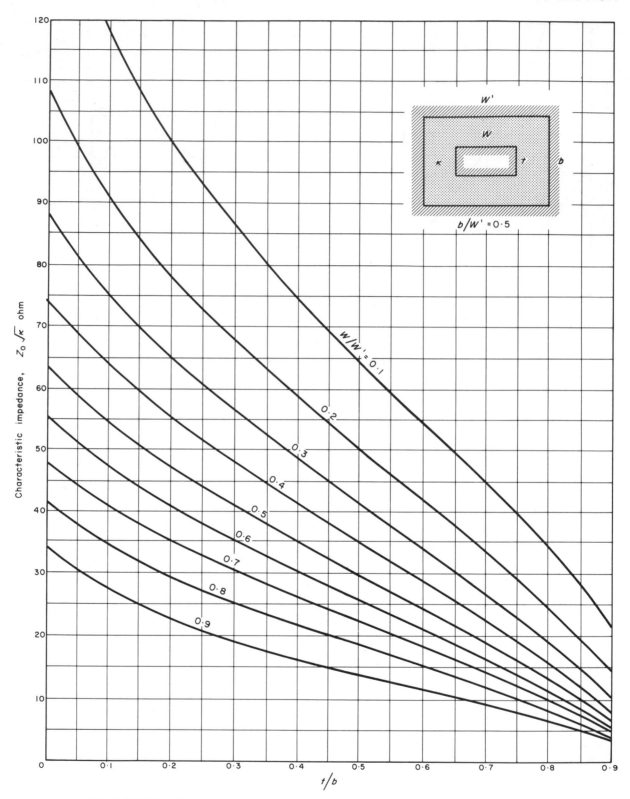

Fig. 3.4(d). The characteristic impedance of the rectangular coaxial line as a function of the cross-sectional dimensions, with $b/W' = 0.5$**. (Courtesy of the I.E.E.)**

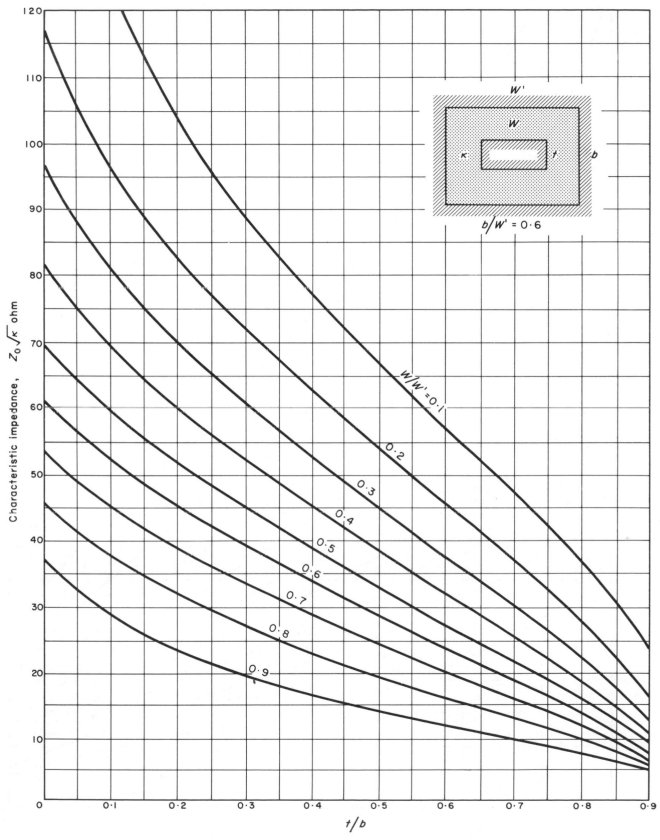

Fig. 3.4(e). The characteristic impedance of the rectangular coaxial line as a function of the cross-sectional dimensions, with $b/W' = 0.6$. (Courtesy of the I.E.E.)

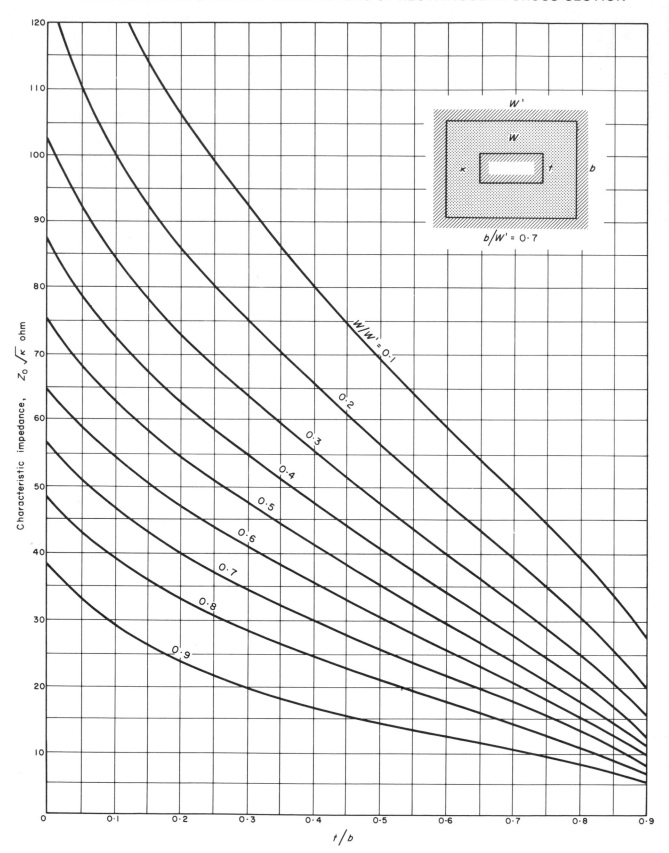

Fig. 3.4(f). The characteristic impedance of the rectangular coaxial line as a function of the cross-sectional dimensions, with $b/W' = 0.7$**. (Courtesy of the I.E.E.)**

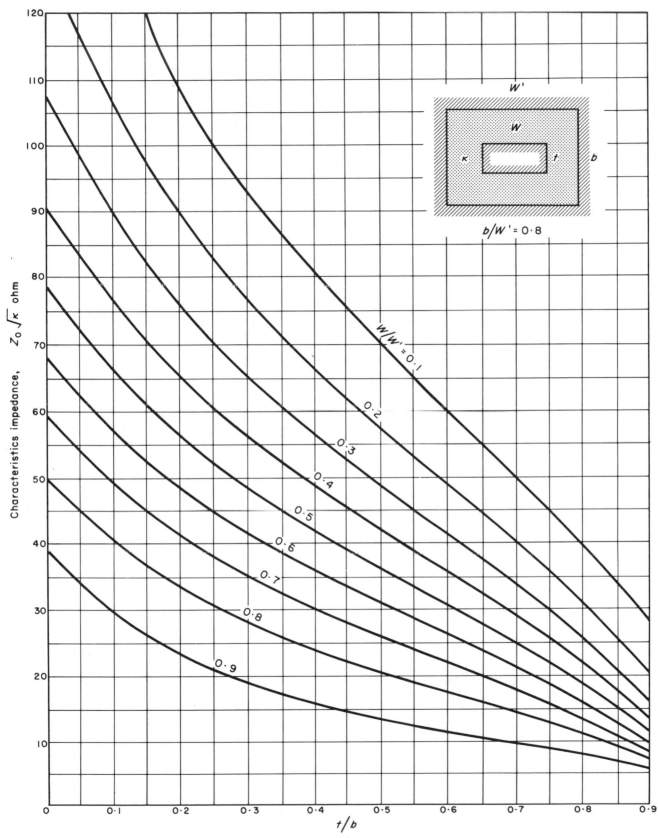

Fig. 3.4(g). The characteristic impedance of the rectangular coaxial line as a function of the cross-sectional dimensions, with $b/W' = 0.8$. (Courtesy of the I.E.E.)

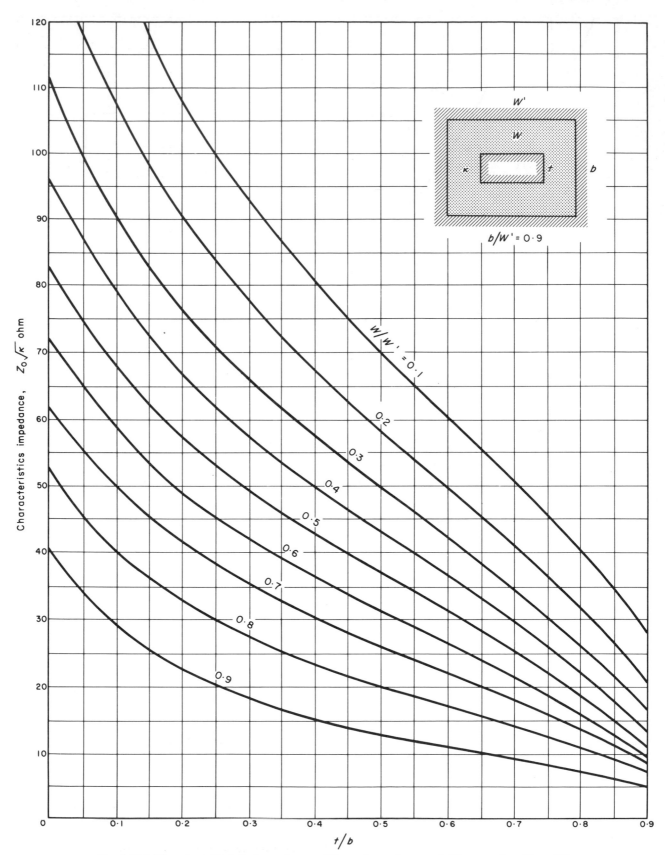

Fig. 3.4(h). The characteristic impedance of the rectangular coaxial line as a function of the cross-sectional dimensions, with $b/W' = 0.9$. **(Courtesy of the I.E.E.)**

$$\frac{K(k_2)}{K'(k_1)} = \frac{W'/b - W/b}{1 - t/b} \qquad (3.4.15)$$

Although of limited value in the general case, Anderson's formulae lead to a much simpler expression for Z_0 in the special case of square coaxial line than do Bowman's. The result is as follows:

$$Z_0\sqrt{\kappa} = 188 \cdot 344 \frac{K(x)}{K'(x)} \text{ ohm} \qquad (3.4.16)$$

where

$$x = \frac{k^2}{1 - k^2} \qquad (3.4.17)$$

and

$$\frac{K(k)}{K'(k)} = \frac{1 - t/b}{1 + t/b}. \qquad (3.4.18)$$

3.4.4 The Rectangular Line with Zero-thickness Centre Conductor

The rectangular coaxial line with zero-thickness centre conductor ($t = 0$ in Fig. 3.3) has been analyzed by Izatt [30], using the Schwarz–Christoffel conformal transformation. This yields an exact expression for the characteristic impedance:

$$Z_0\sqrt{\kappa} = 29 \cdot 976\pi \frac{K(k)}{K'(k)} \text{ ohm} \qquad (3.4.19)$$

where

$$k = \cos\phi \qquad (3.4.20)$$

$$\frac{F(\phi, x)}{K'(x)} = \frac{W}{b} \qquad (3.4.21)$$

and

$$\frac{K(x)}{K'(x)} = \frac{W'}{b} \qquad (3.4.22)$$

The function F is the incomplete elliptic integral of the 1st kind, which is tabulated in [4]. Using these tables in conjunction with the elliptic function formulae given in Section (3.2), Z_0 can be evaluated. However, for certain ranges of parameters, which cover many cases of practical interest, much simpler approximational formulae derived by Izatt [30] can be used. For example, for $Z_0 \leqslant 70$ ohm, $W'/b \geqslant 3/2$

$$Z_0\sqrt{\kappa} \doteqdot \frac{29 \cdot 976\pi b/W}{1 - 2(b/W)L} \text{ ohm} \qquad (3.4.23)$$

where

$$L = \ln\frac{1}{2}\left\{1 - \exp -\pi\left(\frac{W'}{b} - \frac{W}{b}\right)\right\} \qquad (3.4.24)$$

and for $Z_0 \leqslant 70$ ohm, $W'/b \leqslant 2/3$

$$Z_0\sqrt{\kappa} \doteqdot \frac{14 \cdot 988\pi^2}{\ln(8/\pi) - \ln(1 - W/W')} \text{ ohm} \qquad (3.4.25)$$

3.4.5 Other special cases

Using conformal transformations largely based upon Cockcroft's work [21], Chen [22] has obtained expressions for the characteristic impedance of rectangular coaxial lines in which the inner strip conductor is either very thin or very thick. His result for the symmetrical case of an equal gap has already been quoted in Section 3.4.2 (equation 3.4.9). For

$$\frac{(W' - W)}{2} < W \quad \text{and} \quad \frac{(b - t)}{2} < t,$$

Chen gives the following result:

$$Z_0\sqrt{\kappa} = \frac{94 \cdot 172}{\{(t/b)/(W'/b - W/b) + (W/b)/(1 - t/b)\} + C_1 + C_2}$$

$$\text{ohm} \qquad (3.4.26)$$

where

$$C_1 = \frac{1}{\pi}\left\{\ln\left(\frac{g^2 + h^2}{4h^2}\right) + 2\left(\frac{h}{g}\right) \arctan(g/h)\right\}, \qquad (3.4.27)$$

$$C_2 = \frac{1}{\pi}\left\{\ln\left(\frac{g^2 + h^2}{4g^2}\right) + 2\left(\frac{g}{h}\right) \arctan(h/g)\right\}, \qquad (3.4.28)$$

$$g = \tfrac{1}{2}(W' - W) \quad \text{and} \quad h = \tfrac{1}{2}(b - t). \qquad (3.4.29)$$

Conclusion to Section 3.4

For practical design work, the graphs of Figs. 3.4(a)–3.4(h) will usually be found adequate. "Order-of-Magnitude" calculations can quickly be carried out by the use of Bräckelmann's simple formula, equation (3.4.5).

For characteristic impedance formulae applicable to cases where special interrelationships exist between the line dimensions, refer to equations (3.4.8), (3.4.9), (3.4.11), (3.4.19), (3.4.23), (3.4.25), or (3.4.26).

3.5 THE TRIPLATE STRIPLINE

3.5.1 Introduction

This structure, shown in cross-section in Fig. 3.5, was first proposed for use as a transmission line by Barrett and Barnes [23]. Hitherto, it has been

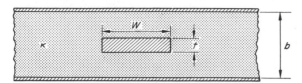

Fig. 3.5. Cross-section of the triplate stripline. κ is the dielectric constant of the medium filling the entire space between the two ground planes

almost universally known as "stripline", since it was the first form of planar strip transmission line to achieve widespread use in microwave technology. However, in view of the now equally extensive use of "micro-" and "High-Q-" striplines it has become necessary to prefix the qualifying term "triplate".

It is readily apparent that the triplate line can be derived from the rectangular coaxial line by the simple step of letting W' become infinite.

Practical stripline circuits are frequently constructed from copper-clad printed-circuit boards, with the result that the centre-conductor thickness "t" is usually very small in comparison with the other transverse dimensions of the line. In any relevant mathematical analyses it is therefore often reasonable to assume that $t = 0$, since this leads to considerable simplification of the problem.

Both situations, i.e. $t = 0$, and $t > 0$, are discussed in the sections which follow.

As in coaxial line systems, it is usually necessary or desirable to ensure the non-propagation of higher-order (non-TEM) modes, and this can be done by ensuring that $b < \lambda/2$. If information about these higher modes is required, it can be deduced from that relevant to rectangular coaxial lines, given in [15–17].

3.5.2 The Zero-thickness Triplate Stripline

When $t/b = 0$ (see Fig. 3.5), an exact derivation of the capacitance per unit length can be per-

formed by the use of conformal transformation. The result is given in [24], and Cohn [25] has used this to obtain the following exact expression for the characteristic impedance of zero-thickness triplate:

$$Z_0\sqrt{\kappa} = 29 \cdot 976\pi \frac{K'(k)}{K'(k)} \text{ ohm} \qquad (3.5.1)$$

where

$$k = \text{sech}\left(\frac{\pi W}{2b}\right). \qquad (3.5.2)$$

Making use of the elliptic integral approximation formulae given in Section 3.2, the characteristic impedance can be expressed in simple closed form thus:

For $W/b \leqslant 0.5$

$$Z_0\sqrt{\kappa} = 29 \cdot 976 \ln 2\left(\frac{1+\sqrt{k}}{1-\sqrt{k}}\right) \text{ohm}. \qquad (3.5.3)$$

For $W/b > 0.5$

$$Z_0\sqrt{\kappa} = \frac{29 \cdot 976\pi^2}{\ln 2\{(1+\sqrt{k'})/(1-\sqrt{k'})\}} \text{ ohm} \qquad (3.5.4)$$

where

$$k' = \tanh\left(\frac{\pi W}{2b}\right). \qquad (3.5.5)$$

These formulae yield results which are virtually exact.

3.5.3 The Finite-Thickness Triplate Stripline

When $t/b > 0$, an exact analysis can still be carried out, using conformal transformation techniques; but with considerable difficulty, and with the disadvantages that the resulting formulae are very complicated and yield the characteristic impedance implicitly rather than explicitly. The problem was first solved by Bates [26] and similar treatments have since been given by Waldron [27, 28] and Ikeda and Sato [29]. Waldron's results are as follows:

$$Z_0\sqrt{\kappa} = 59 \cdot 952 \frac{\pi}{2} \frac{K'(1/\alpha)}{K(1/\alpha)} \text{ ohm} \qquad (3.5.6)$$

where α is related to the cross-sectional dimensions of the line via the following expressions:

$$\frac{t}{b} = \frac{-\{K(\beta') - R\Pi(R', \beta')\}}{R\{\Pi(R', \beta) + \Pi(1 - \alpha^2, \beta') - K(\beta')\}} \quad (3.5.7)$$

$$\frac{W}{b} = \frac{K(\beta) - (1 - \beta^2/\alpha^2)\Pi(\beta^2/\alpha^2, \beta)}{R\{\Pi(R', \beta) + \Pi(1 - \alpha^2, \beta') - K(\beta')\}} \quad (3.5.8)$$

$$R = \frac{\alpha^2 - \beta^2}{\alpha^2 - 1}, \qquad R' = \frac{1 - \beta^2}{1 - \alpha^2}, \qquad \beta' = \sqrt{1 - \beta^2} \quad (3.5.9)$$

and K, Π, are, respectively, complete elliptic integrals of the 1st and 3rd kinds. These equations have been programmed for evaluation by computer, and the resulting data, which are exact, are listed in Table 3.3.

There is an extensive literature on the general topic of triplate stripline, and much of this is devoted to attempts to derive accurate but simple approximation formulae for the characteristic impedance, for easy use by the practical microwave engineer. Probably the best known, and most accurate, of these approximation formulae is that due to Cohn [25]; Chen [22] has given a different formula, but, as has been pointed out by Richardson [31], simple algebraic manipulation shows that the two formulae are in fact identical, and can be written as

$$Z_0\sqrt{\kappa} \doteqdot \frac{94 \cdot 172}{x(W/b) + (1/\pi) \ln F(x)} \text{ ohm} \quad (3.5.10)$$

where

$$F(x) = \frac{(x+1)^{x+1}}{(x-1)^{x-1}} \quad (3.5.11)$$

and

$$x = \frac{1}{1 - t/b}. \quad (3.5.12)$$

Equation (3.5.10) is stated to yield data which are accurate to within 1 per cent for the following parameter ranges:

$$\frac{W}{b} \geqslant 0 \cdot 35\left(1 - \frac{t}{b}\right) \quad (3.5.13)$$

and

$$\frac{t}{b} \leqslant 0 \cdot 25. \quad (3.5.14)$$

An even simpler, and correspondingly less accurate formula is also proposed by Richardson [31]:

$$Z_0\sqrt{\kappa} \doteqdot 29 \cdot 976\pi \ln\left(\frac{1 + W/b}{W/b + t/b}\right) \text{ ohm} \quad (3.5.15)$$

Accuracy of this formula is approximately 1–2 per cent for $W/b \geqslant 1$, 5–6 per cent for $W/b \geqslant 0 \cdot 75$, provided that in both cases $t/b \leqslant 0 \cdot 2$. It is interesting to compare equation (3.5.15) with the almost identical equation (3.4.5) derived by Bräckelmann [14] for the characteristic impedance of rectangular coaxial line.

3.5.4 The Rounded-Edge Triplate Stripline

This version of triplate (see Fig. 3.6 for schematic cross-sectional view) has been devised in order to extend the use of stripline to higher-power operation: the absence of sharp corners on the centre conductor is obviously an essential requirement for such usage, in order to avoid high field-strength areas which could lead to "flashover" and voltage breakdown.

Although graphical data concerning its voltage breakdown properties are available [85], the author is not aware of any published source of data on the characteristic impedance of the rounded-edge stripline. The graphs presented on Fig. 3.6 were plotted from data obtained by numerical solution of implicit relationships between the line dimensions and the characteristic impedance, derived by conformal transformation techniques.†

Note that when $W/b = t/b$, the line cross-section reduces to that of the unshielded slab-line (Section 4.4), giving rise to the limiting boundary indicated on Fig. 3.6. This limiting case can be utilized to develop a simple approximate formula for the characteristic impedance, by a technique similar to that used in connection with the "equal-gap" rectangular coaxial line (see Section 3.4.2). It can be seen from the cross-sectional view on Fig. 3.6 that if the semi-circular edges of the strip are "split off" by vertical diametral planes, the capacitance per unit length of the structure is the same as that of a slab-line composed of the two semi-circular portions, plus the parallel-plate capacitance of the remaining rectangular-section strip: edge-effects due to the corners of the rectangle are removed by the presence of the semi-circular portions. Using this simple model, the characteristic impedance of the rounded-edge

† D. F. Blunden, *private communication*.

TABLE 3.3 THE CHARACTERISTIC IMPEDANCE OF THE TRIPLATE STRIPLINE, EVALUATED FROM WALDRON'S FORMULAE. FOR DEFINITION OF PARAMETERS, SEE FIG. 3.5

$Z_0\sqrt{\kappa}$ ohm	t/b = 0.01 W/b	0.05 W/b	0.10 W/b	0.15 W/b	0.20 W/b	0.25 W/b	0.30 W/b	0.35 W/b
10	8.86585	8.45272	7.95324	7.46362	6.98107	6.50322	6.03098	5.56278
12	7.31234	6.96209	6.54108	6.12975	5.72527	5.32637	4.93233	4.54277
14	6.20245	5.89704	5.53209	5.17682	4.82840	4.48554	4.14756	3.81406
16	5.37003	5.09826	4.77535	4.46212	4.15474	3.85493	3.55899	3.26752
18	4.72260	4.47698	4.18677	3.90624	3.63256	3.36444	3.10120	2.84244
20	4.20465	3.97996	3.71591	3.46153	3.21401	2.97206	2.73497	2.50237
22	3.78087	3.57331	3.33065	3.09768	2.87156	2.65102	2.43534	2.22414
24	3.42773	3.23443	3.00962	2.79448	2.58620	2.38348	2.18564	1.99227
26	3.12891	2.94769	2.73797	2.53792	2.34473	2.15711	1.97436	1.79608
28	2.87278	2.70191	2.50512	2.31801	2.13776	1.96307	1.79325	1.62791
30	2.65081	2.48890	2.30332	2.12743	1.95838	1.79490	1.63630	1.48217
32	2.45658	2.30252	2.12675	1.96066	1.80143	1.64776	1.49897	1.35465
34	2.28519	2.13806	1.97095	1.81352	1.66294	1.51793	1.37779	1.24212
36	2.13286	1.99188	1.83246	1.68272	1.53984	1.40252	1.27008	1.14210
38	1.99656	1.86109	1.70855	1.56570	1.42969	1.29926	1.17370	1.05261
40	1.87388	1.74337	1.59703	1.46037	1.33057	1.20633	1.08696	0.97207
42	1.76290	1.63687	1.49613	1.36508	1.24088	1.12225	1.00849	0.89920
44	1.66200	1.54005	1.40441	1.27845	1.15935	1.04581	0.93715	0.83296
46	1.56987	1.45164	1.32066	1.19936	1.08490	0.97602	0.87201	0.77248
48	1.48543	1.37061	1.24389	1.12686	1.01667	0.91205	0.81231	0.71704
50	1.40774	1.29606	1.17327	1.06016	0.95389	0.85320	0.75739	0.66604
52	1.33603	1.22725	1.10808	0.99859	0.89595	0.79888	0.70669	0.61898
54	1.26963	1.16354	1.04772	0.94159	0.84231	0.74860	0.65976	0.57541
56	1.20798	1.10438	0.99168	0.88867	0.79250	0.70191	0.61620	0.53496
58	1.15059	1.04931	0.93951	0.83940	0.74614	0.65845	0.57565	0.49732
60	1.09703	0.99792	0.89083	0.79343	0.70288	0.61791	0.53782	0.46221
62	1.04693	0.94985	0.84530	0.75043	0.66242	0.57999	0.50245	0.42939
64	0.99997	0.90479	0.80262	0.71014	0.62451	0.54447	0.46932	0.39866
66	0.95587	0.86248	0.76255	0.67230	0.58892	0.51112	0.43822	0.36982
68	0.91438	0.82267	0.72484	0.63671	0.55544	0.47976	0.40899	0.34272
70	0.87526	0.78515	0.68931	0.60317	0.52390	0.45023	0.38146	0.31722
72	0.83834	0.74973	0.65578	0.57153	0.49415	0.42237	0.35551	0.29319
74	0.80343	0.71625	0.62408	0.54162	0.46604	0.39606	0.33102	0.27053
76	0.77038	0.68455	0.59408	0.51332	0.43944	0.37119	0.30787	0.24913
78	0.73905	0.65450	0.56564	0.48650	0.41425	0.34763	0.28598	0.22891
80	0.70931	0.62598	0.53866	0.46106	0.39037	0.32532	0.26525	0.20979
82	0.68104	0.59889	0.51304	0.43691	0.36770	0.30415	0.24560	0.19171
84	0.65415	0.57312	0.48867	0.41396	0.34617	0.28407	0.22698	0.17459
86	0.62854	0.54858	0.46547	0.39212	0.32570	0.26499	0.20932	0.15839
88	0.60413	0.52518	0.44338	0.37132	0.30623	0.24685	0.19256	0.14305
90	0.58084	0.50289	0.42231	0.35151	0.28769	0.22961	0.17665	0.12854
92	0.55860	0.48159	0.40222	0.33262	0.27003	0.21321	0.16155	0.11480
94	0.53735	0.46125	0.38303	0.31460	0.25320	0.19760	0.14722	0.10182
96	0.51702	0.44181	0.36469	0.29740	0.23715	0.18275	0.13361	0.08955
98	0.49757	0.42321	0.34717	0.28097	0.22185	0.16861	0.12070	0.07797
100	0.47894	0.40540	0.33040	0.26527	0.20724	0.15514	0.10845	0.06705
102	0.46110	0.38835	0.31436	0.25027	0.19331	0.14233	0.09684	0.05679
104	0.44398	0.37201	0.29901	0.23592	0.18001	0.13014	0.08584	0.04716
106	0.42757	0.35634	0.28430	0.22220	0.16731	0.11853	0.07543	0.03815
108	0.41181	0.34132	0.27020	0.20907	0.15520	0.10750	0.06560	0.02977
110	0.39668	0.32690	0.25670	0.19651	0.14363	0.09700	0.05632	0.02202
112	0.38215	0.31306	0.24375	0.18448	0.13259	0.08704	0.04759	0.01493
114	0.36818	0.29977	0.23133	0.17298	0.12206	0.07758	0.03939	0.00855
116	0.35475	0.28700	0.21942	0.16197	0.11201	0.06860	0.03172	0.00302
118	0.34184	0.27474	0.20799	0.15143	0.10243	0.06011	0.02459	
120	0.32942	0.26295	0.19703	0.14134	0.09330	0.05208	0.01799	
122	0.31746	0.25161	0.18651	0.13168	0.08460	0.04450	0.01197	
124	0.30595	0.24072	0.17641	0.12244	0.07632	0.03736	0.00657	
126	0.29487	0.23023	0.16672	0.11361	0.06844	0.03067	0.00193	
128	0.28420	0.22015	0.15741	0.10515	0.06100	0.02441		
130	0.27391	0.21045	0.14849	0.09707	0.05385	0.01860		
132	0.26401	0.20112	0.13992	0.08935	0.04712	0.01324		
134	0.25446	0.19213	0.13169	0.08197	0.04075	0.00837		
136	0.24525	0.18349	0.12380	0.07492	0.03474	0.00405		
138	0.23638	0.17517	0.11623	0.06820	0.02908	0.00048		
140	0.22782	0.16716	0.10896	0.06179	0.02377			
142	0.21956	0.15945	0.10199	0.05568	0.01881			
144	0.21160	0.15202	0.09530	0.04987	0.01420			
146	0.20392	0.14487	0.08890	0.04434	0.00996			
148	0.19651	0.13799	0.08275	0.03909	0.00612			
150	0.18936	0.13137	0.07686	0.03412	0.00273			

TABLE 3.3 (continued)

$Z_0\sqrt{\kappa}$ ohm	$t/b = 0.40$ W/b	0.45 W/b	0.50 W/b	0.55 W/b	0.60 W/b	0.65 W/b	0.70 W/b	0.75 W/b
10	5·09894	4·63955	4·18392	3·73285	3·28605	2·84399	2·40718	1·97621
12	4·15746	3·77630	3·39930	3·02655	2·65824	2·29469	1·93633	1·58382
14	3·48480	3·15970	2·83875	2·52206	2·20980	1·90230	1·60000	1·30355
16	2·98031	2·69724	2·41834	2·14368	1·87347	1·60801	1·34775	1·09334
18	2·58792	2·33756	2·09135	1·84939	1·61188	1·37912	1·15155	0·92984
20	2·27401	2·04980	1·82975	1·69315	1·40260	1·19599	0·99459	0·79905
22	2·01718	1·81438	1·61573	1·42134	1·23139	1·04619	0·86619	0·69204
24	1·80315	1·61819	1·43738	1·26082	1·08870	0·92134	0·75917	0·60286
26	1·62206	1·45218	1·28646	1·12499	0·96797	0·81570	0·66862	0·52740
28	1·46682	1·30988	1·15710	1·00857	0·86448	0·72515	0·59101	0·46272
30	1·33229	1·18656	1·04499	0·90767	0·77480	0·64667	0·52324	0·40667
32	1·21458	1·07866	0·94690	0·81939	0·69632	0·57800	0·46488	0·35762
34	1·11071	0·98345	0·86034	0·74149	0·62708	0·51741	0·41295	0·31435
36	1·01838	0·89881	0·78340	0·67224	0·56553	0·46356	0·36679	0·27588
38	0·93578	0·82309	0·71456	0·61029	0·51046	0·41537	0·32549	0·24147
40	0·86143	0·75494	0·65261	0·55453	0·46090	0·37201	0·28833	0·21051
42	0·79417	0·69328	0·59656	0·50409	0·41606	0·33278	0·25471	0·18251
44	0·73302	0·63723	0·54561	0·45823	0·37531	0·29713	0·22417	0·15709
46	0·67719	0·58606	0·49909	0·41637	0·33811	0·26460	0·19631	0·13392
48	0·62602	0·53916	0·45646	0·37801	0·30402	0·23480	0·17081	0·11274
50	0·57895	0·49602	0·41725	0·34274	0·27269	0·20742	0·14740	0·09336
52	0·53551	0·45621	0·38107	0·31020	0·24380	0·18219	0·12586	0·07560
54	0·49531	0·41937	0·34760	0·28010	0·21709	0·15890	0·10603	0·05935
56	0·45798	0·38517	0·31654	0·25219	0·19234	0·13734	0·08775	0·04452
58	0·42326	0·35337	0·28766	0·22626	0·16937	0·11739	0·07092	0·03107
60	0·39087	0·32372	0·26075	0·20211	0·14803	0·09891	0·05545	0·01903
62	0·36061	0·29602	0·23564	0·17961	0·12818	0·08181	0·04129	0·00853
64	0·32228	0·27011	0·21216	0·15860	0·10971	0·06599	0·02844	0·00026
66	0·30571	0·24582	0·19019	0·13899	0·09253	0·05141	0·01693	
68	0·28076	0·22304	0·16961	0·12066	0·07656	0·03804	0·00695	
70	0·25730	0·20164	0·15032	0·10355	0·06176	0·02589		
72	0·23521	0·18153	0·13223	0·08757	0·04809	0·01502		
74	0·21441	0·16262	0·11528	0·07269	0·03553	0·00564		
76	0·19479	0·14482	0·09938	0·05885	0·02409			
78	0·17628	0·12809	0·08451	0·04603	0·01385			
80	0·15882	0·11234	0·07061	0·03423	0·00502			
82	0·14233	0·09754	0·05765	0·02346				
84	0·12678	0·08365	0·04561	0·01378				
86	0·11211	0·07063	0·03448	0·00537				
88	0·09828	0·05845	0·02428					
90	0·08525	0·04709	0·01516					
92	0·07299	0·03655	0·00694					
94	0·06149	0·02683	0·00039					
96	0·05073	0·01796						
98	0·04068	0·01002						

$Z_0\sqrt{\kappa}$ ohm	$t/b = 0.80$ W/b	0.85 W/b	0.90 W/b	0.95 W/b
10	1·55207	1·13629	0·73158	0·34369
12	1·23816	0·90086	0·57463	0·26521
14	1·01394	0·73270	0·46252	0·20916
16	0·84577	0·60657	0·37843	0·16711
18	0·71497	0·50847	0·31303	0·13442
20	0·61034	0·43000	0·26072	0·10826
22	0·52473	0·36579	0·21791	0·08685
24	0·45339	0·31228	0·18224	0·06902
26	0·39302	0·26701	0·15206	0·05393
28	0·34128	0·22820	0·13619	0·04100
30	0·29643	0·19457	0·10377	0·02981
32	0·25719	0·16514	0·08416	0·02007
34	0·22258	0·13918	0·06687	0·01161
36	0·19181	0·11612	0·05154	0·00444
38	0·16429	0·09550	0·03790	
40	0·13954	0·07699	0·02578	
42	0·11718	0·06033	0·01511	
44	0·09690	0·04531	0·00599	
46	0·07848	0·03181		
48	0·06172	0·01979		
50	0·04650	0·00934		
52	0·03274	0·00095		
54	0·02044			
56	0·00969			
58	0·00103			

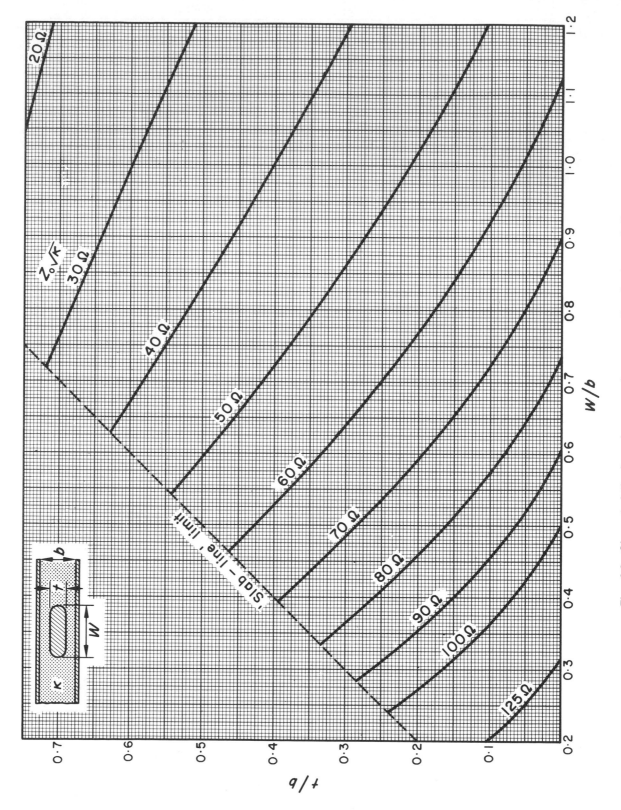

Fig. 3.6. Characteristic impedance, and cross-sectional view, of the "rounded-edge" triplate stripline

stripline is given by

$$Z_0\sqrt{\kappa} \doteqdot \frac{Z_c}{1+4(Z_c/Z)\{(W/b-t/b)/(1-t/b)\}} \text{ ohm} \quad (3.5.16)$$

where $Z_c = Z_0\sqrt{\kappa}$ for the "equivalent" slab-line, i.e. in which $d/b = t/b$ and Z is the characteristic impedance of free space (376·687 ohm).

This expression is virtually exact: its accuracy is almost entirely dependent upon the accuracy of Z_c. Using Frankel's simple approximate formula (see equation 4.4.2), equation (3.5.16) reduces to:

$$Z_0\sqrt{\kappa} \doteqdot \frac{188\cdot344 \ln X}{\pi+2\{(W/b-t/b)/(1-t/b)\}\ln X} \text{ ohm} \quad (3.5.17)$$

where

$$X = \frac{4}{\pi(t/b)} \quad (3.5.18)$$

This formula yields results which are accurate to better than 1 per cent for $t/b \leqslant 0.5$: accuracy falls off rapidly with increasing t/b, but remains within 10 per cent for $0.5 < t/b < 0.7$.

For accurate calculation it is probably best to use equation (3.5.16), inserting values of Z_c obtained from Table 4.3 (Wheeler's results).

Conclusion to Section 3.5

For accurate design work, the data listed in Table 3.4. may be regarded as exact.

For zero-thickness centre conductor, equation (3.5.1) is exact.

For a finite-thickness centre conductor, equation (3.5.10) is highly accurate, provided that $W/b \geqslant 0.35$ $(1-t/b)$ and $t/b \leqslant 0.25$.

For the rounded-edge stripline, the data graphed on Fig. 3.6 although not exact, are sufficiently accurate for all practical purposes: however, it is probably more convenient for design purposes to use equation (3.5.16) in conjunction with Table 4.3.

3.6 THE MICROSTRIP LINE

3.6.1 Introduction

This form of strip transmission line, shown in cross-section in Fig. 3.7, has come into increasing prominence in recent years. Originally conceived in the early 1950's [32, 33], it enjoyed a brief spell of popularity and intensive investigation, but was eventually rejected for microwave use due to the high loss per unit length occasioned by radiation (cf. the single-wire-above-ground discussed in

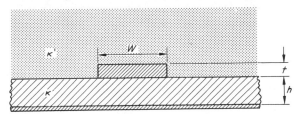

Fig. 3.7. Cross-section of the microstrip line. The medium of dielectric constant κ' is assumed to be infinite in extent. Usually, $\kappa \gg \kappa'$, and in most practical cases $\kappa' = 1$ (air)

Chapter 2). This was largely a result of the low dielectric constants (around 2) of the substrate materials then in use. Further developments were prevented by the lack of availability both of high-dielectric-constant, low-loss materials, and of suitable methods of processing and production. In more recent years, these difficulties have been completely overcome, and this, coupled with the ever-increasing demands for miniaturized microwave circuitry for use in weapons, aerospace and satellite applications, has led to renewed intensity of interest in microstrip. Further stimulus for the re-investigation of microstrip circuits has come from the somewhat unexpected area of high-speed electronic computers: these involve the use of nanosecond pulse techniques, which in turn obviously require the use of circuitry appropriate to the GHz range of frequencies.

The strip thicknesses used in present-day microstrip circuits are quite small; between 0·0003 and 0·003 in in the majority of cases: this is largely a consequence of the fact that the manufacturing processes involved were originally developed for low-frequency printed-circuit boards, or for thin- and thick-film microelectronic circuits. Hence, as in the case of triplate stripline, the zero-thickness strip approximation is often sufficiently accurate for practical design purposes.

The existing volume of literature on the topic of microstrip is of truly encyclopaedic proportions, and is still increasing. Even a very condensed review would occupy far more space than is permissible, or appropriate, within the present text; therefore, the paragraphs which follow will be limited to brief discussion of those contributions of greatest practical use for basic microwave circuit design and analysis.

The fact that propagation takes place in a non-homogeneous dielectric medium (i.e. more than one dielectric is present) means that a true TEM mode of propagation cannot exist, since the dielectric boundary conditions cannot be satisfied by such a mode. However, the field configuration of the actual mode of propagation closely resembles the TEM structure, and in the majority of cases of practical interest the "pseudo-TEM" mode assumption is well justified.

Because of the boundary–condition difficulty just mentioned, most of the earlier analyses assumed a single uniform dielectric medium as well as zero-thickness strip, and took the actual dielectric structure into account by the use of an "effective" dielectric constant κ_e, derived in various ways. This approach was also adopted in several of the more recent analyses, but in addition many investigators have resorted to numerical techniques in order to take account of both the finite-thickness and two-dielectric cases.

For convenient discussion, the various analyses to be described will be classified under four subheadings, according to the "mathematical model" assumed in each case. For clarity, the structure illustrated in Fig. 3.7 will be referred to as "microstrip", and the single-dielectric version (where $\kappa' = \kappa$) will be referred to as "single-dielectric microstrip".

3.6.2 The Zero-thickness Single-dielectric Microstrip Line

The first analysis of this structure (see Fig. 3.7, putting $\kappa' = \kappa$ and $t = 0$) was given by Assadourian and Rimai [33] using conformal transformation techniques, but in view of the limited accuracy, and restrictive conditions under which it is valid, the results obtained are really only of historical interest. An almost identical treatment was later given by Kovalyev [34]; but a much more accurate and useful analysis, also using conformal transformations, was presented by Wheeler [35], and his results are as follows. For $W/h < 2$

$$Z_0\sqrt{\kappa'} \doteq 376{\cdot}687\left\{\frac{1}{2\pi}\ln\left(\frac{8h}{W}\right)+\frac{1}{16\pi}\left(\frac{W}{2h}\right)^2-\cdots\right\}\text{ohm} \quad (3.6.1)$$

and for $W/h > 2$

$$Z_0\sqrt{\kappa'}\doteq\frac{376{\cdot}687}{(W/4h)+(1/2\pi)\ln\{17{\cdot}08(W/2h+0{\cdot}92)\}}\text{ohm} \quad (3.6.2)$$

where κ' is the dielectric constant of the uniform medium.

Hilberg [36], with the aim of producing simple formulae of fair accuracy for a wide range of parameter values, has also applied conformal mapping techniques to the problem, and obtained the following results. For $0 \leqslant h/W \leqslant 1{\cdot}5$

$$Z_0\sqrt{\kappa'}\doteq\frac{188{\cdot}344\pi}{\cosh^{-1}(F)}\text{ohm} \quad (3.6.3)$$

where the function F is given implicitly by

$$\frac{\pi W}{2h}=F-\ln F-1. \quad (3.6.4)$$

For $1{\cdot}5 \leqslant h/W \leqslant \infty$

$$Z_0\sqrt{\kappa'}\doteq\frac{188{\cdot}344}{\pi}\ln\frac{8h}{W}\text{ohm} \quad (3.6.5)$$

Although these formulae give results in close agreement with Wheeler's, over the range $0{\cdot}3 \leqslant W/h \leqslant 30$, their practical value is somewhat limited by the implicit nature of equation (3.6.4), and in ease of application they appear to offer no advantage over the use of Wheeler's more complex, but explicit, formulae. Note, in fact, that the right-hand side of equation (3.6.5) is the first term of Wheeler's formula (equation 3.6.1).

The *exact* analysis of zero-thickness single-dielectric microstrip has been presented by Schneider [37]: the formulae are

$$Z_0\sqrt{\kappa'} = 188{\cdot}344\frac{K'(k)}{K(k)}\text{ohm} \quad (3.6.6.)$$

$$\text{dn}^2(2\alpha K(k)) = \frac{E(k)}{K(k)} \quad (3.6.7)$$

$$\frac{W}{h} = \frac{2}{\pi}\frac{\partial}{\partial\alpha}\ln\theta_4\left(\alpha,\frac{K'}{K}\right) \quad (3.6.8)$$

where $E(k)$ is the complete elliptic integral of the 2nd kind, dn is the Jacobian elliptic function, θ_4 is the theta function, and κ' is the dielectric constant of the uniform medium.

These formulae provide implicit relationships between Z_0 and the line parameters, but because of their complexity, they are difficult, if not impossible, to solve without computer assistance. Schneider has therefore given the following approximations, with a claimed accuracy of $\pm 0{\cdot}25$ per cent for $0 \leqslant W/h \leqslant 10$. For $W/h \leqslant 1$

$$Z_0\sqrt{\kappa'} = 59{\cdot}952\ln\left(\frac{8h}{W}+\frac{W}{4h}\right)\text{ohm} \quad (3.6.9)$$

For $W/h \geqslant 1$

$$Z_0\sqrt{\kappa'} = \frac{119\cdot904\pi}{(W/h)+2\cdot42-0\cdot44(h/W)+\{1-(h/W)\}^6}\ \text{ohm}$$

(3.6.10)

If $W/h > 10$, the accuracy obtained by the use of equation (3.6.10) is plus or minus 1 per cent.

The single-dielectric form of microstrip has comparatively few practical applications, because

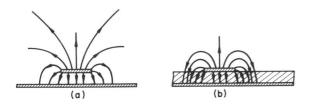

(a)   (b)

Fig. 3.8. Schematic illustration of the electric field distribution in two forms of microstrip. Note the "field-concentrating" effect of the high-dielectric-constant substrate in (b)

of its high radiation loss. The reason for this will be obvious if one compares the schematic electric field patterns displayed in Figs. 3.8(a) and 3.8(b): the "field-concentrating" effect of two-dielectric microstrip is readily apparent. Nevertheless, the analysis of single-dielectric microstrip is of considerable interest because the results obtained are of use in the analysis of the "real" microstrip of Fig. 3.7.

3.6.3   The Finite-thickness Single-dielectric Microstrip Line

For many applications, particularly in the computer-circuitry area, the assumption of zero strip-thickness is not valid, and several authors, including Assadourian and Rimai [33], Wheeler [38], and more recently Kondratyev et al. [39], have endeavoured to produce reasonably simple methods of including this factor, or correcting for it.

As in the zero-thickness case, Wheeler's formulae seem to offer the best combination of simplicity with accuracy: equations (3.6.1) and (3.6.2) remain valid provided that an "effective width", W_e, is substituted for the actual width W, where

$$W_e = W + \frac{t}{\pi}\left(1+\ln\frac{2x}{t}\right)$$

(3.6.11)

where $x = h$ for $W > (h/2\pi) > 2t$, and $x = 2\pi W$ for $(h/2\pi) > W > 2t$.

Kondratyev's results, although derived from a very accurate conformal transformation, and hence virtually exact, cannot, unfortunately, be described as "reasonably simple", as they involve complex implicit relationships between various elliptic integrals and hyperbolic functions. Since the single-dielectric microstrip is of such limited practical interest, it will not be discussed further.

3.6.4   The Zero-thickness Microstrip Line

As in the previous two instances, Wheeler's comprehensive treatments [35, 38] have provided reasonably simple, yet very accurate, closed formulae for the characteristic impedance of zero-thickness microstrip. For narrow strips, $W/h < 1$

$$Z_0 = \frac{376\cdot687}{\pi\sqrt{2}\sqrt{\kappa+1}}\left\{\ln\left(\frac{8h}{W}\right)+\frac{1}{32}\left(\frac{W}{h}\right)^2-\frac{1}{2}\left(\frac{\kappa-1}{\kappa+1}\right)\left(\ln\frac{\pi}{2}+\frac{1}{\kappa}\ln\frac{4}{\pi}\right)\right\}$$

ohm.   (3.6.12)

For wide strips, $W/h > 1$

$$Z_0 = \frac{376\cdot687}{2\sqrt{\kappa}}\left[\frac{W}{2h}+0\cdot441+0\cdot082\left(\frac{\kappa-1}{\kappa^2}\right)\right.$$
$$\left.+\left(\frac{\kappa+1}{2\pi\kappa}\right)\left\{1\cdot451+\ln\left(\frac{W}{2h}+0\cdot94\right)\right\}\right]^{-1}\ \text{ohm}$$

(3.6.13)

In these formulae, κ is the dielectric constant of the substrate supporting the strip. Good agreement with experimental measurements (using non-zero, but very small thickness, so that $t/h \not> 0\cdot005$) is obtained for $2 < \kappa < 10$ and $0\cdot1 < W/h < 5$.

The rather simpler approximate formulae derived by Schneider [37] for the single-dielectric case (equations (3.6.9) and (3.6.10) above) can be extended to cover the present case simply by replacing the "uniform" dielectric constant κ' by the "effective" dielectric constant κ_e. Wheeler [38] has given graphs from which this can be determined, for a given set of line parameters, via a "filling fraction", q, defined by

$$\kappa_e = 1 + q(\kappa - 1) \qquad (3.6.14)$$

where κ is the dielectric constant of the substrate. Similar graphs are given by Presser [40]. However, by use of curve-fitting techniques, Schneider [37] has derived a satisfactorily accurate empirical formula for κ_e, as follows

$$\kappa_e = \frac{\kappa+1}{2} + \frac{\kappa-1}{2}\left(1 + \frac{10h}{W}\right)^{-1/2}. \qquad (3.6.15)$$

It is claimed that this formula yields values for $\sqrt{\kappa_e}$ within 1 per cent of Wheeler's results, over the entire range $0 \leqslant W/h < \infty$ and $1 \leqslant \kappa < \infty$.

Apart from the various analytic and empirical approaches which have been employed in the analysis of zero-thickness microstrip, a number of treatments involving wholly or partly numerical methods have been presented. Notable among these, for their importance in treating the case of finite thickness strip, are the works of Silvester [41], Bryant and Weiss [42] and Hill, Reckord, and Winner [43]: these will be discussed in Section 3.6.5.

3.6.5 The Finite-thickness Microstrip Line

This is the real "practical" structure illustrated in Fig. 3.7, which has come to be of such great importance both in microwave device technology and, more recently, in computer circuitry. As a result of this, the finite-thickness microstrip has probably received more attention than any of the more specialized versions described in the previous paragraphs, and some of the many empirical, analytic, and numerical approaches used in its analysis are reviewed below.

The earliest accurate analysis of finite-thickness microstrip was given by Wheeler [38]: his zero-thickness formulae given above (equations (3.6.12) and (3.6.13)) remain valid for the finite-thickness case, provided that the actual strip width W is replaced by the effective strip width W_e, given by equation (3.6.11).

Note particularly that Wheeler states that these formulae are valid only for $\kappa_e = \kappa = 1$: for $\kappa > 1$ he proposed a different W_e. However, Escritt† states (on the basis of a detailed experimental and theoretical investigation), that the use of formula (3.6.11) leads to the closest agreement

† P. A. Escritt, *private communications.*

with experiment over the widest range of parameters.

For computer circuit applications, which generally involve dielectric constants of between 2 and 6, a very simple and quite accurate formula has been derived by Kaupp [44], on the basis of numerous experimental measurements of the dielectric constants of various substrate materials. This makes use of the simple formula for the characteristic impedance of a "single-wire-above-ground" (equation (2.4.3)), combined with a well-known "equivalence" between circular and rectangular cross-sections [45, 46], to yield

$$Z_0\sqrt{\kappa_e} = 59{\cdot}952 \ln (4h/d) \text{ ohm} \qquad (3.6.16)$$

where

$$\kappa_e = 0{\cdot}475\kappa + 0{\cdot}67 \qquad (3.6.17)$$

and

$$d = 0{\cdot}536W + 0{\cdot}67t. \qquad (3.6.18)$$

For parameter values in the ranges $W/h < 1{\cdot}25$ and $0{\cdot}1 < t/W < 0{\cdot}8$, the above formulae yield results which are within 5 per cent or so of experimentally-determined values (provided that $2{\cdot}5 < \kappa < 6$).

Silvester's analysis of zero-thickness microstrip has already been mentioned in Section 3.6.4, and its application to the case of finite-thickness microstrip is also treated in [41]. The method combines the classical "method of images" with the use of the very powerful Green's function techniques [47] to determine the potentials and charges of a system of N "elementary sub-strips", into which the actual strip is divided. Although this is, in principle, a purely "analytic" method, in order to obtain specific results it is necessary to resort to numerical techniques: for example, the potential of the kth sub-strip is given by an expression of the form:

$$V_k = \sum_j^N \left\{ \frac{1}{w_k} \cdot \frac{1}{w_j} \int_{w_k} \int_{w_j} G(x, y : x', y')\, dx\, dy\, dx'\, dy' \right\} q_j$$

where w_k, w_j are the widths of the sub-strips k and j, G is the appropriate Green's function, and q_j is the charge per unit length on the jth sub-strip. Numerical techniques are obviously required here for evaluation of the integrals (because of the complexity of G), for carrying out the summations (the accuracy of the results will obviously depend upon the smallness of w_k and w_j), and for inversion of the resulting large-order matrix equation of the form

$$\mathbf{V} = \mathbf{PQ}$$

where $\mathbf{V}$ and $\mathbf{Q}$ are column matrices with elements V_k and q_j respectively.

The great merit of Silvester's method is that it can be extended to many other similar types of line, in particular a pair of coupled lines (as will be seen in Chapter 6) of zero or finite thickness, and involving two or more dielectric layers if desired. The main difficulty in applying the method is in deriving the appropriate Green's function; but Silvester has given explicit formulae for zero-thickness and finite-thickness microstrip, and these have been utilized in a computer program for the evaluation of Z_0. Fundamentally, this uses the technique outlined above to evaluate the capacitance of the microstrip structure both with and without dielectrics present, and hence evaluates Z_0 and $\sqrt{\kappa_e}$. Some useful results are displayed on Figs. 3.9–3.11, and in Table 3.4. These have been widely used, and their accuracy confirmed, in experimental projects; and they may for all practical purposes be regarded as virtually exact. Fig. 3.9 presents a "universal" curve for zero-thickness microstrip characteristic impedance, applicable to any form of microstrip for which the *effective* dielectric constant κ_e is known: some values of κ_e for substrate dielectric constants of 3·78 and 9·6 are given in Table 3.4. Fig. 3.10 gives graphs of microstrip characteristic impedance versus W/h for $\kappa = 3\cdot78$ and 9·6, with strip thickness as parameter, and Fig. 3.11 shows the variation of characteristic impedance with strip thickness for some typical cases.

As was mentioned in Section 3.5.4, similar analyses involving the use of Green's functions and numerical techniques have been presented by Bryant and Weiss [42] and by Hill *et al.* [43]. Bryant and Weiss' work was originally reported in [48], and this gives a more detailed description of the analysis, and presents extensive tables of impedance of zero-thickness microstrip for several values of κ.

The tremendous interest in microstrip circuits has led to the situation where numerous independent groups of investigators have been simultaneously working on similar problems, with the somewhat unfortunate result that a spate of papers has appeared, all of which deal with basically the same topics, namely the evaluation of microstrip impedances and propagation parameters. A comprehensive bibliography would in itself constitute a volume of appreciable size, and here we shall mention only a few of the many valuable contributions which have appeared: this does not, of course, carry any implication that those mentioned are in any way superior to the many others not referred to.

Yamashita and Mittra [49] have reported a variational analysis which yields results in good agreement with those obtained by other methods and by experimental measurement. Similar results have been obtained by Stinehelfer [50], by the use of finite-difference techniques. Judd *et al.* [51], in seeking to reduce the very lengthy computation times required for finite-difference methods, have presented as an alternative an analytic method based upon the use of "ortho-normal" block analysis; which has yielded accurate (experimentally verified) values of microstrip impedance. All relevant equations are detailed in [51].

3.6.6 The True Nature of Propagating Modes in Microstrip

It has already been mentioned that the presence of more than one dielectric in the microstrip structure imposes boundary conditions which cannot be satisfied by a purely TEM mode. Although the actual field structure closely resembles that which would be appropriate to a simple TEM mode, there exist other points of difference between the "real" and "assumed" modes, the most important of which is probably the dispersive nature of the "real" mode. A true TEM mode, if it could exist, would be non-dispersive, i.e. its propagation velocity would be constant, and independent of frequency. This is not true of the microstrip propagation mode, and the effects of this can be quite serious at the higher frequencies of operation, in particular when designing circuit components which involve quarter- or half-wavelengths, or in designing computer circuits, in which time-domain behaviour is of the utmost importance.

The topic of the determination of the complete propagating mode spectrum of microstrip has therefore received considerable attention. It was first qualitatively discussed by Deschamps [52],

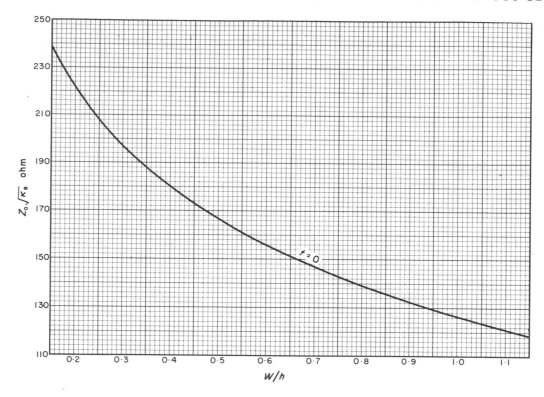

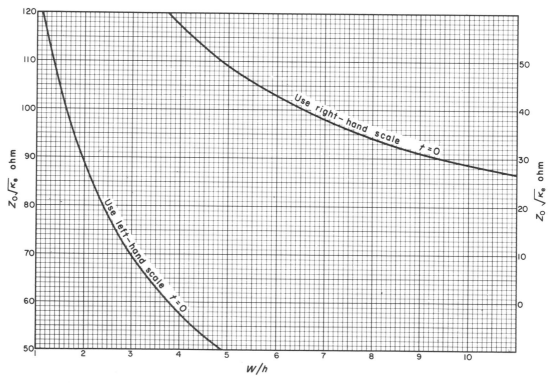

Fig. 3.9. Universal curve for the characteristic impedance of zero-thickness microstrip. κ_e is the effective dielectric constant of the microstrip structure

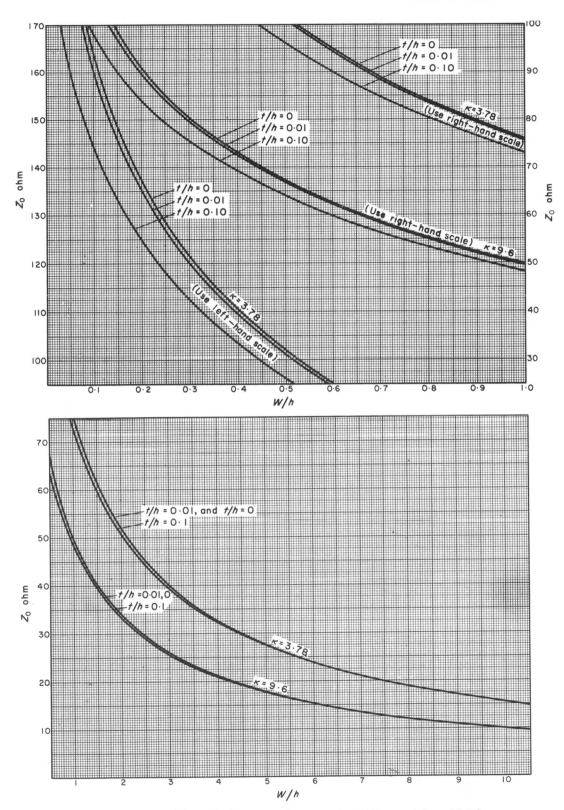

Fig. 3.10. The characteristic impedance of finite-thickness microstrip, for three values of strip thickness, for substrate dielectric constants $\kappa = 3.78$ (e.g. quartz) and $\kappa = 9.6$ (e.g. alumina)

TABLE 3.4   THE CHARACTERISTIC IMPEDANCE Z_0 AND EFFECTIVE DIELECTRIC CONSTANT κ_e FOR FINITE-THICKNESS MICROSTRIP, WITH $\kappa = 3.78$ AND $\kappa = 9.6$

| | $\kappa = 3.78$ | | | | | | $\kappa = 9.6$ | | | | | |
| | $t/h = 0$ | | $t/h = 0.01$ | | $t/h = 0.10$ | | $t/h = 0$ | | $t/h = 0.01$ | | $t/h = 0.10$ | |
W/h	Z_0 ohm	κ_e	Z_0 ohm	κ_e	Z_0 ohm	κ_e	Z_0 ohm	κ_e	Z_0 ohm	κ_e	Z_0 ohm	κ_e
0·100	164·10	2·57	160·05	2·52	144·02	2·36	109·03	5·82	106·80	5·67	97·88	5·11
0·125	155·43	2·58	152·10	2·54	138·05	2·39	103·21	5·85	101·37	5·72	93·57	5·19
0·150	148·36	2·59	145·52	2·55	133·01	2·41	98·46	5·88	96·90	5·76	89·96	5·26
0·175	142·37	2·60	139·92	2·56	128·63	2·43	94·44	5·90	93·09	5·79	86·83	5·33
0·200	137·20	2·61	135·03	2·58	124·74	2·44	90·96	5·93	89·77	5·83	84·08	5·38
0·225	132·63	2·61	130·70	2·59	121·25	2·46	87·90	5·95	86·83	5·86	81·60	5·43
0·250	128·56	2·62	126·82	2·59	118·06	2·47	85·16	5·97	84·20	5·88	79·36	5·47
0·275	124·87	2·63	123·29	2·60	115·14	2·49	82·68	5·99	81·81	5·91	77·31	5·51
0·300	121·51	2·63	120·06	2·61	112·44	2·50	80·42	6·01	79·63	5·94	75·42	5·55
0·325	118·42	2·64	117·09	2·62	109·93	2·51	78·35	6·03	77·62	5·96	73·67	5·59
0·375	112·90	2·65	111·77	2·63	105·39	2·53	74·64	6·07	74·02	6·00	70·50	5·65
0·425	108·09	2·66	107·11	2·65	101·36	2·55	71·41	6·11	70·87	6·05	67·71	5·71
0·475	103·82	2·68	102·97	2·66	97·74	2·57	68·55	6·14	68·08	6·09	65·20	5·77
0·500	101·86	2·68	101·07	2·67	96·06	2·57	67·23	6·16	66·79	6·10	64·04	5·79
0·550	98·23	2·69	97·52	2·68	92·92	2·59	64·79	6·19	64·40	6·14	61·87	5·84
0·600	94·92	2·70	94·30	2·69	90·04	2·60	62·57	6·22	62·22	6·18	59·89	5·89
0·650	91·89	2·71	91·34	2·70	87·38	2·62	60·53	6·25	60·23	6·21	58·06	5·93
0·700	89·10	2·72	88·60	2·71	84·91	2·63	58·66	6·28	58·39	6·24	56·36	5·97
0·750	86·51	2·73	86·06	2·72	82·61	2·64	56·92	6·31	56·68	6·28	54·79	6·01
0·800	84·10	2·74	83·70	2·73	80·46	2·66	55·31	6·34	55·09	6·31	53·31	6·05
0·850	81·85	2·75	81·49	2·74	78·44	2·67	53·80	6·37	53·60	6·34	51·93	6·09
0·900	79·74	2·76	79·41	2·75	76·53	2·68	52·38	6·40	52·20	6·37	50·63	6·13
0·950	77·75	2·77	77·45	2·76	74·73	2·69	51·05	6·42	50·88	6·40	49·40	6·16
1·00	75·87	2·78	75·61	2·77	73·03	2·70	49·79	6·45	49·64	6·42	48·23	6·19
1·25	67·86	2·82	67·69	2·81	65·68	2·75	44·43	6·58	44·34	6·56	43·24	6·35
1·50	61·53	2·86	61·42	2·85	59·80	2·80	40·19	6·69	40·13	6·68	39·25	6·49
1·75	56·35	2·89	56·29	2·89	54·95	2·84	36·74	6·80	36·71	6·79	35·98	6·62
2·00	52·04	2·92	51·99	2·92	50·88	2·87	33·87	6·90	33·84	6·89	30·24	6·73
2·25	48·37	2·95	48·34	2·95	47·40	2·91	31·43	6·99	31·41	6·99	30·90	6·84
2·50	45·20	2·98	45·18	2·98	44·39	2·94	29·33	7·08	29·32	7·07	28·89	6·94
2·75	42·45	3·01	42·43	3·00	41·75	2·97	27·50	7·16	27·49	7·15	27·12	7·03
3·00	40·02	3·03	40·00	3·03	39·42	2·99	25·90	7·23	25·89	7·23	25·57	7·11
3·25	37·86	3·05	37·85	3·05	37·34	3·02	24·47	7·30	24·46	7·30	24·19	7·19
3·50	35·93	2·07	35·91	3·07	35·48	3·04	23·20	7·37	23·19	7·37	22·95	7·26
4·00	32·62	3·11	32·60	3·11	32·27	3·08	21·02	7·49	21·01	7·48	20·83	7·40
4·50	29·88	3·14	29·85	3·14	29·61	3·12	19·22	7·60	19·21	7·59	19·08	7·51
5·00	27·57	3·17	27·54	3·17	27·36	3·15	17·71	7·69	17·70	7·68	17·60	7·62
5·50	25·60	3·20	25·56	3·20	25·43	3·18	16·43	7·78	16·40	7·77	16·33	7·72
6·00	23·90	3·23	23·85	3·22	23·76	3·21	15·31	7·86	15·29	7·84	15·24	7·80
6·50	22·41	3·25	22·35	3·25	22·29	3·24	14·34	7·93	14·31	7·91	14·28	7·88

TABLE 3.4 (continued)

	$\kappa = 3.78$						$\kappa = 9.6$					
	$t/h = 0$		$t/h = 0.01$		$t/h = 0.10$		$t/h = 0$		$t/h = 0.01$		$t/h = 0.10$	
W/h	Z_0 ohm	κ_e	Z_0 ohm	κ_e	Z_0 ohm	κ_e	Z_0 ohm	κ_e	Z_0 ohm	κ_e	Z_0 ohm	κ_e
7.00	21.09	3.27	21.03	3.27	21.00	3.26	13.49	8.00	13.45	7.98	13.44	7.96
7.50	19.92	3.29	19.85	3.28	19.84	3.28	12.73	8.06	12.69	8.03	12.69	8.03
8.00	18.88	3.31	18.80	3.30	18.81	3.30	12.05	8.12	12.01	8.09	12.01	8.09
8.50	17.93	3.33	17.85	3.32	17.88	3.32	11.44	8.18	11.40	8.14	11.41	8.15
9.00	17.08	3.34	16.99	3.33	17.04	3.34	10.89	8.23	10.84	8.18	10.86	8.20
9.50	16.31	3.36	16.21	3.34	16.27	3.35	10.39	8.28	10.34	8.23	10.37	8.25
10.00	15.60	3.37	15.50	3.36	15.56	3.37	9.93	8.32	9.88	8.27	9.91	8.30

and later by Wu [53]. Other more qualitative treatments, and experimental data, have been published by Hartwig et al. [54], Eaves and Bolle [55], and Napoli and Hughes [56]. Apart from dispersive effects, the propagation loss in microstrip is also of great interest and importance, and in this area important contributions have been made by Welch and Pratt [57], Pucel et al. [58], Hartwig et al. [59], Schneider [60], and Kitamura et al. [61].

3.6.7 Shielded Microstrip or "Microstrip-in-a-box"

The majority of microstrip circuits in practical use will be located within containers, racks, or cabinets, and hence, either intentionally or unintentionally, will have conducting boundaries other than ground-planes located at non-infinite distances from the strips. These will obviously affect the circuit behaviour, to a greater or lesser degree depending upon the distances between the microstrip circuit and the conducting boundaries, and hence considerable attention has been given to the analysis of the "screened microstrip" structure illustrated in cross-section in Fig. 3.12. The problem is obviously even more complex than in the case of "plain" microstrip, so that purely analytic solutions are virtually impossible to obtain. Numerical analyses, based upon the "pseudo-TEM" mode assumption, have been given by a number of authors. The earliest relevant treatment seems to have been published by Guckel [62], for the simple case of air-filled zero-thickness microstrip with a "top-plate" but no side walls (this can obviously, with equal justification, be regarded as "unsymmetrical triplate": but for present purposes this viewpoint yields no

advantages). From Guckel's graph it can be seen that if the top-plate is located at a distance Nh above the strip, where h is the height of the strip above the ground plane and N is an integer, the characteristic impedance Z_0 becomes independent of N for $N > 10$. In view of the "field-concentrating" effect produced by the presence of the high dielectric constant substrate in "real" microstrip, it may hence be inferred (with some caution, however; bearing in mind the possible presence of "waveguide" modes) that the top-plate will have little effect on circuit behaviour provided that $N \gg 1$.

This conclusion is supported by the results of Yamashita's variational analysis of the zero-thickness microstrip with top-plate [79]: it is shown that, with a sapphire substrate ($\kappa = 9.9$), both the characteristic impedance and the propagation wavelength are virtually independent of N for $N > 5$. See also Fig. 3.13, which shows typical results calculated by the use of Bräckelmann's method (see below, and [63]).

In most practical circuits it is usual to ensure that N is fairly large, and also that any side-walls are located at adequately large distance from the strip.

A method of obtaining more precise data on the effects of side and top walls has been developed by Bräckelmann [63]. This consists of a very general analysis of a rectangular section strip within a rectangular conducting shield, containing up to three different dielectric layers; and the axes of strip and shield need not be coincident.

Although, having formulated the basic equations, Bräckelmann quickly specializes to the simple case (i.e. coincident axes, one dielectric) of symmetrical rectangular coaxial line (see Section 3.4), the fundamental analysis can ob-

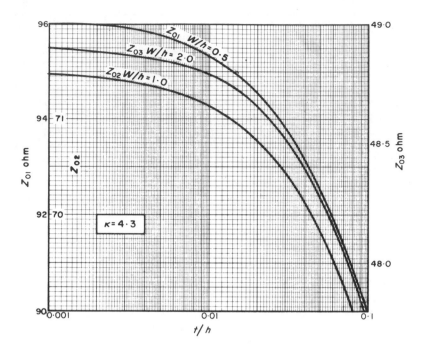

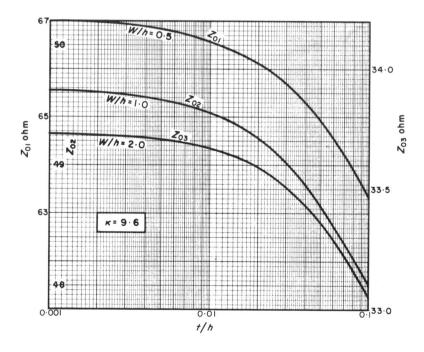

Fig. 3.11. Illustration of the effects of strip thickness on microstrip characteristic impedance in some typical cases

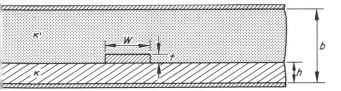

Fig. 3.12. Cross-section of "screened" microstrip. κ **is the dielectric constant of the substrate, and** κ' **is the dielectric constant of the medium filling the remaining space between the two ground planes. Usually,** $\kappa \gg \kappa'$

viously be adapted (though with tedium and difficulty), to many of the other transmission lines already discussed, including "microstrip-in-a-box" (see Fig. 3.14). In particular, the cut-off frequencies of the higher-order modes can be obtained, as well as the TEM mode characteristic impedance. In a later publication [64], Bräckelmann extends the analysis to two or more coupled strips within a rectangular shield.

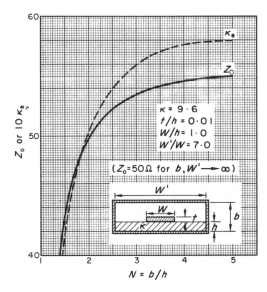

Fig. 3.13. Illustration of the dependence of the characteristic impedance of screened microstrip on the height (Nh) of the upper ground-plane (top-plate) above the substrate, for the particular case $\kappa = 9 \cdot 6$

Other (graphical) data concerning characteristic impedance, propagation constant, and loss per wavelength are given in Stinehelfer's paper [50]. More recently, Mittra and Itoh [73] have presented a semi-rigorous analysis of multilayer shielded microstrip structures, based on function-theoretic techniques, which yields accurate results for the characteristic impedances of such structures. The method is extremely complex and it is rather

unfortunate that the authors restrict themselves to a limited graphical display of the valuable results obtained.

Conclusion to Section 3.6

(a) *Microstrip*. The data given in Table 3.5 and Figs. 3.9 and 3.10 can be regarded as exact, for all practical purposes. In order to deal with substrates of other dielectric constants, the use of zero-thickness strip data will yield quite accurate results. For example, use Fig. 3.9 in conjunction with Schneider's formula for κ_e (equation (3.6.15)): or use Wheeler's zero-thickness formulae (equations (3.6.12) and (3.6.13)) with the "effective width" given by equation (3.6.11).

For simplicity with reasonable accuracy, use Kaupp's formulae (3.6.16)–(3.6.18), *but only* for the range of dielectric constants quoted.

(b) *Screened Microstrip*. In the case of screened microstrip, either with or without side walls, the number of independent parameters is so large

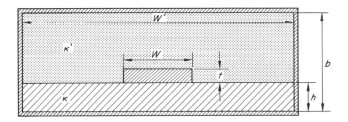

Fig. 3.14. Cross-section of "microstrip-in-a-box", i.e. screened microstrip with side walls

that it is virtually impossible to produce sufficient tables or graphs to cover even the most commonly-used combinations. For accurate work, therefore, the reader has little choice other than to consult the references listed, and if necessary prepare his own computerized calculation procedure.

For less stringent requirements, much of the data pertaining to "ordinary" microstrip can be used, *provided* that a suitable expression or value for κ_e can be derived, empirically or otherwise.

3.7 MICROSTRIP "DERIVATIVES"

3.7.1 Introduction

The widespread success of microstrip circuitry has led to fairly intensive investigation of similar line structures, and at the time of writing several different, but related, versions are known, but as yet unproven. Hence, for purposes of reference, it has been thought desirable to include a few of the more promising of these, even though little detailed information is available at present.

3.7.2 "Suspended-substrate" Microstrip

Two versions of this have been proposed, both of which are illustrated in cross-section in Fig. 3.15. As is apparent, they both incorporate a strip conductor supported upon a thin dielectric substrate, the under-surface of which is unmetallized. The two structures differ only in the shape of the surrounding conductor, which is circular in one case and rectangular in the other.

The circular version was proposed, and put to practical use, by Okean [65]. Under the classification system adopted in this book, this line should perhaps more properly be treated in the next chapter, but its obvious close relationship to "ordinary" microstrip is sufficient reason for its

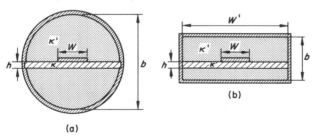

(a)   (b)

Fig. 3.15. Cross-sections of two alternative versions of suspended-substrate microstrip.

inclusion here. But see Section 4.6 for a similar transmission line, which constitutes the "limiting case" of the one at present under discussion.

An exact analysis of this structure is not known, but under the pseudo-TEM mode assumption, the characteristic impedance is given by

$$Z_0\sqrt{\kappa_e} = \frac{3335 \cdot 64}{C_0} \text{ ohm} \qquad (3.7.1)$$

where κ_e is the *effective* dielectric constant of the structure given by:

$$\kappa_e = \frac{C}{C_0} \qquad (3.7.2)$$

C is the capacitance per unit length (pF/m) of the structure and C_0 is the capacitance per unit length (pF/m) in the absence of dielectric.

An approximate analysis of the structure, using conformal transformation and the method of images (see [65] for details) yields:

$$C \doteqdot 17 \cdot 72 \left\{ \frac{K(k)}{K'(k)} + \tfrac{1}{2}(\kappa - \kappa') \frac{K(k_1)}{K'(k_1)} \right\} \text{picofarads/metre} \quad (3.7.3)$$

where

$$\sqrt{1 - k^2} = \frac{(1 - W/b)^2 + (h/b)^2}{(1 + W/b)^2 + (h/b)^2} \qquad (3.7.4)$$

and

$$\sqrt{1 - k_1^2} = \frac{\tanh X}{\tanh Y} \qquad (3.7.5)$$

with

$$X = \frac{\pi b}{8h} \left(1 + \frac{h^2}{b^2} \right)(1 - k^2)^{1/4} \qquad (3.7.6)$$

and

$$Y = \frac{\pi b}{8h} \left(1 + \frac{h^2}{b^2} \right)(1 - k^2)^{-1/4} \qquad (3.7.7)$$

Although fairly crude approximations are involved in the analysis, graphs presented by Okean [65] for the case $\kappa' = 1$ seem to indicate that agreement between theoretical and experimental results is quite close, with a maximum error in characteristic impedance of around 5 per cent. In view of the limited utilization of this version of microstrip, and the rather large number of parameters involved, it has not been considered worthwhile to include any detailed numerical data. This limited usage is perhaps surprising, in view of its several apparent advantages: like the rectangular version (discussed below) it offers low dielectric loss, and possibly cheaper manufacturing cost (since no metallization is needed on the underside of the substrate); and in addition it would appear to be more "compatible" with ordinary coaxial line than any other version of microstrip. The author has no personal experience in the use of this particular line, but it may well be that the advantages mentioned are more than outweighed by the fairly obvious difficulties of manufacturing such a structure to close and repeatable tolerances.

For those wishing to pursue the matter further, design data can fairly readily be obtained from equations (3.7.1)–(3.7.7) by the use of the elliptic integral formulae given in Section 3.2 (equations (3.2.5)–(3.2.10)). See also Section 4.6.

The rectangular version of suspended-substrate microstrip (Fig. 3.15(b)) is rather more popular, though not yet attracting the same intensive investigation and development as "ordinary" microstrip. The main advantages claimed over the latter are less degradation of performance due to variations in the substrate dielectric constant, lower dielectric loss, and easier realization of high impedance lines (because these are wider, due to the reduced "effective" dielectric constant engendered by the air layer below the substrate). Fairly detailed accounts of the finite-difference analysis of "shielded" suspended-substrate microstrip (i.e. with side and top walls present) have been given by Brenner [80, 81]. Reference [81] in particular details all finite-difference formulae, and includes a print-out of the FORTRAN program used on an IBM 7094 computer. Several graphs of characteristic impedance and propagation velocity as functions of the other line parameters are given, [80, 81], but only for glazed alumina substrates ($\kappa = 8 \cdot 0$).

Similar graphs, in some cases verified by experimental data, have been given by Yamashita [79] both for microstrip with a top-plate, and for "unshielded" suspended-substrate microstrip (i.e. no side walls). In the latter case, the substrates concerned are polystyrene ($\kappa = 2 \cdot 55$) and quartz ($\kappa = 4 \cdot 5$): the calculated data are derived from a variational analysis of the structure. The same method has been applied [82] to the derivation of some graphs of W/h versus κ for constant Z_0 (50 ohm and 75 ohm), relevant to the suspended-substrate microstrip with a top-plate. This structure can also be treated as a special case of triplate stripline involving three dielectric layers, for which the analytical equations are detailed by Mittra and Itoh [73], but these need numerical solution for particular cases. Presentation of data is limited to a few graphs relevant to "ordinary" microstrip with a top-plate (see Fig. 3.12).

A similar structure, but with the strip on the under side instead of the top side of the substrate, has been qualitatively discussed by Schneider [37].

The types of line discussed in this section can be designed by the use of the data and formulae given in Sections 3.6.2–3.6.7, *provided that* suitable values of κ_e can be arrived at: in this connection, Schneider [37] has provided some relevant graphical data.

For approximate design purposes, simple "electrostatic" formulae for κ_e are sometimes adequate.

3.7.3 The "Slot" Line

This form of transmission line was first proposed by S. B. Cohn [66], as an alternative to microstrip, and in fact, it can legitimately be regarded as the waveguide dual of microstrip, since it can be derived from the latter by replacing metal—dielectric interfaces by dielectric—dielectric interfaces, and vice-versa; as can be seen from Fig. 3.16. Its claimed advantages are simplicity of manufacture, higher attainable values of Z_0, low loss (provided that $\kappa > 10$), and a propagating field structure which has elliptically-polarized regions that can conveniently be utilized to produce non-reciprocal ferrite devices. In addition, it is readily compatible with microstrip circuits on the same substrate.

The main disadvantage of slot-line is its dispersive nature; this renders the "pseudo-TEM" mode assumption much less justifiable than in previous cases discussed, and in particular the "characteristic impedance" is quite strongly frequency-dependent, though this dependence becomes less marked for small values of W/h.

Cohn has given a fairly detailed analysis of the slot-line [67], and produces "first-order" and "second-order" approximations for propagation wavelength, characteristic impedance, etc.: but Mariani *et al.* [68] have shown that although the

Fig. 3.16. Cross-section of slot-line

theoretical and measured values of wavelength are in close agreement (within 2 per cent), measured values of characteristic impedance can be 30 per cent or more below those predicted by theory.

For preliminary "order-of-magnitude" calculations, a fairly simple approximate expression for the "pseudo-TEM" mode characteristic impedance of slot-line can be derived from the electrostatic capacitance C of the structure, using the well-known formula

$$Z_0 = \frac{1}{vC} \text{ ohm} \qquad (3.7.8)$$

where v is the velocity of propagation. C is given by Wolfe [69] as follows:

$$C = \varepsilon_0 \frac{\kappa}{2} \frac{K'(k)}{K(k)} \text{ farads/metre} \qquad (3.7.9)$$

where

$$k = \tanh\left(\frac{\pi W}{4h}\right) \qquad (3.7.10)$$

and ε_0 is the permittivity of free space. Since $v = f\lambda_e$, where λ_e is the effective propagation wavelength and f = frequency, and since Cohn [67] gives a simple approximate formula for λ_e, namely

$$\lambda_e = \lambda\sqrt{\frac{2}{1+\kappa}} \qquad (3.7.11)$$

then by substitution in equation (3.7.8) and rearrangement one can derive

$$Z_0 \doteq 533\frac{\sqrt{1+\kappa}}{\kappa}\frac{K(k)}{K'(k)} \text{ ohm} \qquad (3.7.12)$$

κ is the dielectric constant of the substrate. For small values of W/h, this formula, although independent of frequency, yields results in reasonable agreement with Cohn's, but in view of their very poor accuracy as judged by comparison with experiment, the formula is only of value for "order of magnitude" estimation, as mentioned above.

Useful data on slot-line ferrite devices has been reported by Robinson and Allen [70].

The slot-line concept has been further developed by Eaves and Bolle [71]; they give a detailed analysis of "shielded slot-line", which is simply slot-line with conducting boundary planes located above and below it, and with the entire structure filled with a *single* homogeneous dielectric, as shown in Fig. 3.17. In a similar, but less strict,

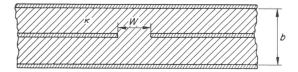

Fig. 3.17. Cross-section of shielded slot-line.

sense to that in which slot-line was regarded as the dual of microstrip, the shielded slot-line can be regarded as the dual of triplate stripline (see Section 3.5). However, since Eaves and Bolle state that any useful modes of propagation are non-TEM, this "line" will not be discussed further.

3.7.4 The Coplanar Waveguide

This structure, shown in cross-section in Fig. 3.18, forms yet another variation on the microstrip theme, and was devised by Wen [72]. Claimed

advantages are virtually the same as for slot-line, including ready adaptation to the design of non-reciprocal (ferrite) devices, and it is noteworthy that, unlike slot-line, coplanar waveguide can with reasonable justification be regarded as a "pseudo-TEM" transmission line. A quasi-static approximate analysis can be carried out, similar to, but more justifiable than, that used for slot-line, which assumes the dielectric substrate to be infinitely thick, and, via conformal transformation, leads to the following formulae:

$$Z_0\sqrt{\kappa+1} \doteq 133\cdot2\frac{K'(k)}{K(k)} \text{ ohm} \qquad (3.7.13)$$

where

$$k = (1+2g/W)^{-1} \qquad (3.7.14)$$

The effect of the non-infinite substrate thickness decreases with increasing κ: Wen [72] has shown that equation (3.7.13) yields results in good, though not exact, agreement with experiment for $\kappa = 9\cdot5$, 16, and 130.

3.7.5. The Microstrip Trough Line

This structure, shown in cross-section on Fig. 3.19, bears a remote resemblance to the trough

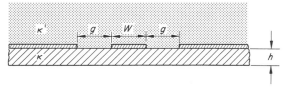

Fig. 3.18. Cross-section of coplanar waveguide

line discussed in Section 4.5, and was proposed by Fritsche [86] under the name of "chamber-line" (*Kammerleitung*). It is claimed to combine the advantages of triplate stripline (presumably those appertaining to any shielded structure) with those of microstrip (open structure, ease of manufacture). Approximate formulae for the line parameters are derived under the "pseudo-TEM" mode assumption, and are as follows (a zero thickness strip is assumed):

$$Z_0\sqrt{\kappa_e} \doteq 59\cdot952 \ln\left\{\frac{W'}{W}\cdot\phi\left(\frac{h}{b},\frac{W'}{b}\right)\right\} \text{ohm} \qquad (3.7.15)$$

where the effective dielectric constant κ_e is given by

$$\kappa_e \doteq \kappa(1-\tfrac{2}{9}\phi)+\tfrac{2}{9}\phi \qquad (3.7.16)$$

and ϕ is a function of the line dimensions.

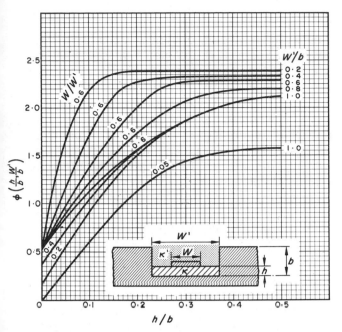

Fig. 3.19. Graphical representation of the ϕ function required in the evaluation of the characteristic impedance of microstrip trough line; and schematic cross-section of this line. (Graphs redrawn from those given by Fritsche in [86])

The nature of ϕ is left unspecified, but graphs of ϕ versus h/b, as a function of W'/b, are given. These have been redrawn, and are displayed on Fig. 3.19.

3.8 THE "HIGH-Q" TRIPLATE STRIPLINE

3.8.1 Introduction

From the cross-sectional diagram shown in Fig. 3.20 it is apparent that one may regard this line

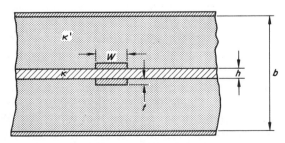

Fig. 3.20. Cross-section of the High-Q triplate stripline

either as a microstrip derivative (if one locates a "mirror-reflection" plane midway between the two strips), or, perhaps, as a triplate derivative (regarding it as a "split-strip" version of triplate).

Neither viewpoint, unfortunately, leads to any significant simplification of the analysis problem, but as in so many of the previous instances, the pseudo-TEM mode assumption is reasonably valid, and on this basis the structure can, in principle, be analyzed either by conformal transformation or by numerical techniques.

This form of stripline developed virtually in parallel with triplate and microstrip; although not quite enjoying the widespread popularity of the former (nor suffering the temporary rejection of the latter!) it has nevertheless come into very extensive use, probably because it is so easy and cheap to produce from double-clad printed-circuit boards of standard type by standard methods.

It was originally thought that this configuration of conductors and dielectric would yield much lower loss than the "standard" dielectric-filled triplate, because of the low proportion of dielectric included in the propagation area (in the most common version, where $\kappa' = 1$). However, this expectation was not entirely fulfilled, because although it is true that only a small proportion of electric field line lengths are "immersed" in dielectric, it is unfortunately the case that the dielectric-filled region adjacent to the strips is a high-field-strength region. Thus losses are still far from negligible, although certainly less than in dielectric-filled triplate.

The first (and, until recently, only) detailed analysis of High-Q triplate was given by Foster [74] and he treated the problem in two parts: first, the determination of the characteristic impedance of what he called the "basic" line, i.e. the structure of Fig. 3.20 with $\kappa' = \kappa = 1$, and $t = 0$; and then the evaluation of the propagation velocity v, from which one can derive an effective dielectric constant κ_e.

The same course will be adopted in the paragraphs which follow.

3.8.2 The Characteristic Impedance of High-Q Triplate

Denoting the characteristic impedance of the basic (i.e. completely air-filled) line by Z_b, to avoid later confusion, Foster's result [74] is:

$$Z_b = 29 \cdot 976\pi \frac{K'(x)}{K(x)} \text{ ohm} \qquad (3.8.1)$$

Unfortunately, the formulae relating x to the

cross-sectional dimensions of the line are extremely complex (and implicit) functions of elliptic integrals (complete and incomplete) and Jacobian theta- and zeta-functions: closely similar formulae, apparently independently derived, have been presented by Yamamoto *et al.* [76] and by Sato and Ikeda (contributing to Matsumoto's work) [12]. For most practical applications it is simpler to make use of alternative expressions derived by Cohn [75], which though not exact, are nearly so.

Cohn has analyzed the basic line, but from the viewpoint that the two strips are separate and different, though coupled, transmission lines; and he has derived expressions for the even- and odd-mode characteristic impedances of this pair of coupled lines. Referring to the definitions of these parameters given in Chapter 1 (Section 1.5) it is evident that the even-mode impedance of Cohn's structure is identical with $2Z_b$. Hence, his formulae can be used directly, with the result that Z_b is still given by equation (3.8.1), but x is now given by a much simpler, though still implicit, function of the line parameters:

$$\frac{\pi W}{2b} = \tanh^{-1}\sqrt{X} - \frac{h}{b}\tanh^{-1}\frac{\sqrt{X}}{x} \qquad (3.8.2)$$

where

$$X = \frac{x(\dot{x} - h/b)}{(1 - xh/b)}. \qquad (3.8.3)$$

Equations (3.8.1)–(3.8.3) are virtually exact, provided that $W/h > 0.35$, which is usually true in practice.

If, in addition to the above condition, the following restriction is also obeyed (which is frequently the case)

$$\frac{W/b}{1 - h/b} \geqslant 0.35 \qquad (3.8.4)$$

then the following explicit formulae can be used, with good accuracy:

$$Z_b = \frac{29.976\pi(1 - h/b)}{W/b + 0.4413(1 - h/b) - C/\pi} \text{ ohm} \qquad (3.8.5)$$

where

$$C = \left(1 - \frac{h}{b}\right)\ln\left(1 - \frac{h}{b}\right) + \frac{h}{b}\ln\left(\frac{h}{b}\right). \qquad (3.8.6)$$

Having obtained Z_b, it is next necessary to determine the propagation velocity, v, appropriate to the structure when required dielectric media are present. The characteristic impedance of High-Q triplate is then given by

$$Z_0 = Z_b\frac{v}{c} \qquad (3.8.7)$$

where c is the velocity of light *in vacuo* (2.997925×10^8 m/s). Foster [74] obtained values of v by the use of a variational technique, which has the obvious disadvantage of requiring lengthy computations for any particular combination of dimensions and dielectric constants. However, his published results cover many cases of practical interest, and are reproduced in Fig. 3.21, together with experimental results which provide adequate verification of the computed curves.

An alternative approach to the evaluation of Z_0 is provided by the use of numerical techniques. In common with the variational technique just mentioned, this suffers from the disadvantage of requiring lengthy computations, but this is to a large extent outweighed by the advantage of obtaining directly, from a single computational process, values of Z_b, Z_0, and v, instead of (as in the previous case) having to combine the results with those of an almost equally lengthy analytical process in order to obtain usable data.

The numerical analysis of the High-Q triplate structure, or any similar structure (e.g. triplate or microstrip) can most conveniently be carried out if the conducting boundaries are continuous, i.e. if side walls are present, so that the structure analyzed should more accurately be designated "shielded High-Q triplate". This is discussed in the following section.

3.8.3 The Shielded High-Q Triplate Line

This structure is the "High-Q" analogue of rectangular coax, or microstrip-in-a-box, and hence needs no illustration, being the same as the structure of Fig. 3.20 with the addition of a pair of symmetrically-disposed side walls, separated by a distance W'.

No analytical treatment is known, but numerical analyses of particular cases have been given by Green and Pyle [77] and by Earle and Benedek [78]: both treatments are confined to the case where $2.55 \leqslant \kappa \leqslant 2.65$ (e.g. Rexolite 2200), and the side wall separation W' is such that $W/W' \leqslant 0.35$.

The data presented in [77] are in the form of graphs of Z_0 and v versus W, for $t = 1/16$ in and $b = 5/16$ in, whereas a rather more extensive display in [78] presents Z_0 and v versus W/b for a range of t/b.

However, Gish and Graham [83] have given a much more detailed, and more accurate, semi-

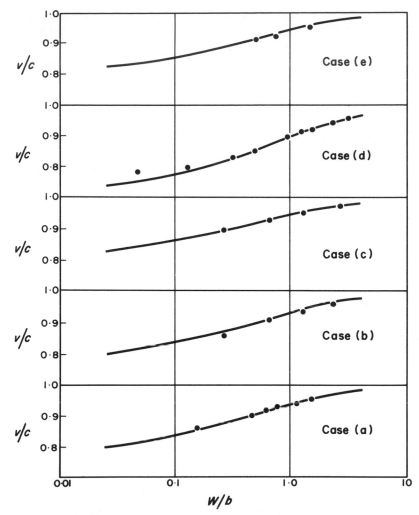

Fig. 3.21. Velocity ratios in the High-Q triplate stripline as functions of W/b
(see Fig. 3.20 for definition of parameters).
Case (a): Silicone-resin-bonded fibreglass substrate. $\kappa = 4\cdot16$, $\kappa' = 1$, $h/b =$
0·125 (at 3 GHz); Case (b): Resin-bonded paper substrate. $\kappa = 3\cdot76$, $\kappa' = 1$,
$h/b = 0\cdot17$ **(at 1 GHz); Case (c): Teflon-bonded fibreglass substrate.** $\kappa = 3\cdot16$,
$\kappa' = 1$, $h/b = 0\cdot17$ **(at 1 GHz); Case (d): Silicone-resin-bonded fibreglass**
substrate. $\kappa = 4\cdot16$, $\kappa' = 1$, $h/b = 0\cdot25$ **(at 9 GHz); and Case (e): Polythene**
substrate. $\kappa = 2\cdot25$, $\kappa' = 1$, $h/b = 0\cdot50$ **(at 3 GHz) (Reproduced from [74] by**
kind permission of the I.E.R.E.)

analytic treatment, which involves the setting-up of a variational expression using Green's functions [47]. The Rayleigh–Ritz method is then applied to the calculation of Z_0 and v. Detailed large-scale graphs of these parameters as functions of W/b are given, for $0\cdot05 \leqslant t/b \leqslant 0\cdot5$, and for $1 \leqslant W'/b \leqslant 4$, but only for $\kappa' = 1$, $\kappa = 2\cdot35$ (Duroid 5870). It is pointed out that this is a much more economical method of analysis as compared with Green and Pyle's finite-difference technique [77], being some seven times faster in terms of computation time. Results obtained have been confirmed by experimental measurement, and the analysis is given in sufficient detail for the reader to prepare his own computer-programmed procedure if desired.

However, data of sufficient accuracy for many design requirements can probably be obtained by the use of Gish and Graham's extensive graphs in conjunction with Foster's data obtained for other dielectric constants (see Fig. 3.21): these can be used to obtain an empirical "correction factor" for use with certain of Gish and Graham's results.

Finally, the topic of higher-order mode propagation in shielded High-Q triplate has been treated in some detail, by Pregla and Schlosser [84]. Using a field-component matching technique, the necessary equations are derived analytically, and then solved by numerical techniques. The cut-off frequencies of the E_{11}, H_{10}, and H_{01} modes are evaluated, for a wide range of values of h/b, W'/W, W/b and t/b, and for three dielectric constants: $\kappa = 2\cdot32$ (Rexolene S), $\kappa = 2\cdot6$ (Custom Poly QR), and $\kappa = 9\cdot0$ (Alumina).

These data are displayed on numerous, though small-scale, graphs.

REFERENCES

1. Schelkunoff, S. A. *Applied Mathematics for Engineers and Scientists.*
   Chapter XIV. 2nd edn, Van Nostrand, Princeton, N.J. (1965).
2. Cunningham, J. *Complex Variable Methods.*
   Chapter 4. Van Nostrand, Lond. (1965).
3. Bowman, F. *Introduction to Elliptic Functions.*
   Dover, New York (1961).
4. Abramowitz, M. and Stegun, I. A. *Handbook of Mathematical Functions.*
   Applied Mathematics Series No. 55. National Bureau of Standards, Washington, D.C. (1965).
5. Hilberg, W. "From Approximations to Exact Relations for Characteristic Impedances", *Trans. I.E.E.E.*
   MTT-17, pp. 259–265 (May 1969).
6. Whittaker, E. T. and Watson, G. N. *Modern Analysis.*
   4th edn, p. 486. Cambridge University Press (1940).
7. Bowman, F. *Proc. Lond. math. Soc.*
   39, 211 (1935); **41**, 271 (1936).
8. Conning, S. W. "The Characteristic Impedance of Square Coaxial Line", *Trans. I.E.E.E.*
   MTT-12, p. 468 (July, 1964).
9. Green, H. E., in *Advances in Microwaves.*
   Vol. 2, p. 350, Ed. Leo Young, Academic Press, New York (1967).
10. Anderson, G. M. "The Calculation of the Capacitance of Coaxial Cylinders of Rectangular Cross-Section", *Trans. A.I.E.E.*
    69, pt II, 728–731 (1950).
11. Green, H. E. "The Characteristic Impedance of Square Coaxial Line", *Trans. I.E.E.E.*
    MTT-11, pp. 554–555 (Nov. 1963).
12. Matsumoto, A. (Ed.). "Microwave Filters and Circuits".
    Supplement 1 to *Advances in Microwaves.*
    Ch. V, Academic Press, New York (1970).
13. Lin, W-G and Chung, S-L. "A New Method of Calculating the Characteristic Impedances of Transmission Lines".
    Acta phys. sin.
    19, 249–257 (April 1963) (in Chinese).
14. Bräckelmann, W. "Wellentypen auf der Streifenleitung mit rechteckigem Schirm", *A.E.Ü.*
    21, 641–648 (Dec. 1967).
15. Bräckelmann, W. *et al.* "Die Grenzfrequenzen von höheren Eigenwellen in Streifenleitungen", *A.E.Ü.*
    21, 112–120 (March, 1967).
16. Baier, W. "Wellentypen in Leitungen aus Leitern rechteckigen Querschnitts", *A.E.Ü.*
    22, 179–185 (April, 1968).
17. Baier, W. "The Calculation of TEM, TE, and TM-Waves in Shielded Strip Transmission Lines", *Digest of the I.E.E.E. G-MTT Symposium.*
    pp. 21–31 (1968).
18. Skiles, J. J. and Higgins, T. J. "Determination of the Characteristic Impedance of UHF Coaxial Rectangular Transmission Lines", *Proc. natn. electron. Conf. (Chicago).*
    10, 97–108 (1954).
19. Schneider, M. V. "Computation of Impedance and Attenuation of TEM-Lines by Finite Difference Methods", *Trans. I.E.E.E.*
    MTT-13, pp. 793–800 (Nov. 1965).
20. Metcalf, W. S. "Characteristic impedance of rectangular transmission lines", *Proc. I.E.E.*
    112, 2033–2039 (Nov., 1965).
21. Cockcroft, J. D. "The Effect of Curved Boundaries on the distribution of Electrical Stress round Conductors", *J. I.E.E.*
    26, 385–409 (1927–28).
22. Chen, T-S. "Determination of the Capacitance, Inductance and Characteristic Impedance of Rectangular Lines", *Trans. I.R.E.*
    MTT-8, pp. 510–519 (Sept., 1960).
23. Barrett, R. M. and Barnes, M. H. "Microwave Printed Circuits", *Radio TV News.*
    46, 16 (1951).
24. Oberhettinger, F. and Magnus, W. *Anwendung der Elliptischen Funktionen in Physik und Technik.*
    Springer-Verlag (1949).
25. Cohn, S. B. "Characteristic Impedance of Shielded-Strip Transmission Line", *Trans. I.R.E.*
    MTT-2, pp. 52–55 (July, 1954).
26. Bates, R. H. T. "Characteristic Impedance of the Shielded Slab Line", *Trans. I.R.E.*
    MTT-4, pp. 28–33 (Jan., 1956).
27. Waldron, R. A. *Theory of Guided Electromagnetic Waves.*
    pp. 187–191, Van Nostrand Reinhold, Lond. (1970).
28. Waldron, R. A. "Theory of a strip-line cavity for measurement of dielectric constants and gyromagnetic-resonance line-widths", *Trans. I.E.E.E.*
    MTT-12, pp. 123–131 (1964).
29. Ikeda, T. and Sato, R. "Analysis of the Characteristic Impedance and the Effective Resistance of Strip Transmission Lines with a Rectangular Inner Conductor by the use of Conformal Mapping", *Electronics Communs Jap.*
    50, 111–117 (March, 1967) (*see also* [12]).
30. Izatt, J. B. "Characteristic Impedance of two special forms of transmission line", *Proc. I.E.E.*
    111, 1551–1555 (Sept., 1964).
31. Richardson, J. K. "An Approximate Formula for Calculating Z_0 of a Symmetric Strip Line", *Trans. I.E.E.E.*
    MTT-15, pp. 130–131 (Feb., 1967).
32. Greig, D. D. and Engelmann, H. F. "Microstrip—A New Transmission Technique for the Kilomegacycle Range", *Proc. I.R.E.*
    40, 1644–1650 (Dec., 1952).
33. Assadourian, F. and Rimai, E. "Simplified Theory of Microstrip Transmission Systems", *Proc. I.R.E.*
    40, 1651–1657 (Dec., 1952).
34. Kovalyev, I. S. "Calculation of the fields of non-symmetrical stripline taking into account the thickness of the current-carrying strip", *Izvestiya Vishykh Uchebnikh Zavedenii Energetika.*
    pp. 105–109 (in Russian) (1965).
35. Wheeler, H. A. "Transmission Line Properties of Parallel Wide Strips by a Conformal-Mapping Approximation", *Trans. I.E.E.E.*
    MTT-12, pp. 280–289 (May, 1964).
36. Hilberg, W. "Eine einfache und gute Näherung fur den

Wellenwiderstand paralleler Streifenleitungen", *A.E.Ü.*
22, 122–126 (March, 1968).

37. Schneider, M. V. "Microstrip Lines for Microwave Integrated Circuits", *B.S.T.J.*
48, 1421–1444 (May–June, 1969).

38. Wheeler, H. A. "Transmission-Line Properties of Parallel Strips Separated by a Dielectric Sheet", *Trans. I.E.E.E.*
MTT-13, pp. 172–185 (March, 1965).

39. Kondratyev, B. V. *et al.* "Investigation of a Nonsymmetric Strip Line with a Charged Slab of Arbitrary Thickness", *Radio Engng electron. Phys.*
14, 449–452 (March, 1969).

40. Presser, A. "RF Properties of Microstrip Line", *Microwaves.*
7, 53–55 (March, 1968).

41. Silvester, P. "TEM wave properties of microstrip transmission lines", *Proc. I.E.E.*
115, 43–48 (Jan., 1968).

42. Bryant, T. G. and Weiss, J. A. "Parameters of Microstrip Transmission Lines and of Coupled Pairs of Microstrip Lines", *Trans. I.E.E.E.*
MTT-16, pp. 1021–1027 (Dec., 1968).

43. Hill, Y. M. *et al.* "A General Method for obtaining Impedance and Coupling Characteristics of Practical Microstrip and Triplate Transmission Line Configurations", *IBM Jl Res. Dev.*
13, 314–322 (May, 1969).

44. Kaupp, H. "Characteristics of Microstrip Transmission Lines", *Trans. I.E.E.E.*
EC-16, pp. 185–193 (April, 1967).

45. Springfield, W. K. "Designing Transmission Lines into Multilayer Boards", *Electronics.*
38, 90–108 (Nov. 1st, 1965).

46. Cohn, S. B. "Problems in Strip Transmission Lines", *Trans. I.R.E.*
MTT-3, pp. 119–126 (March, 1955).

47. Roach, G. F. *Green's Functions—Introductory theory with applications.*
Van Nostrand Reinhold, Lond. (1970).

48. Bryant, T. J. and Weiss, J. A. "Parameters of Microstrip Transmission Lines and of Coupled Pairs of Microstrip Lines", *Worcester Polytechnic Institute Report.*
(July 15th, 1968).

49. Yamashita, E. and Mittra, R. "Variational Method for the Analysis of Microstrip Lines", *Trans. I.E.E.E.*
MTT-16, pp. 251–256 (April, 1968).

50. Stinehelfer, H. E. "An Accurate Calculation of Uniform Microstrip Transmission Lines", *Trans. I.E.E.E.*
MTT-16, pp. 439–444 (July, 1968).

51. Judd, S. V. *et al.* "An Analytical Method for Calculating Microstrip Transmission Line Parameters", *Trans. I.E.E.E.*
MTT-18, pp. 78–87 (Feb., 1970).

52. Deschamps, G. A. "Theoretical Aspects of Microstrip Waveguides", *Trans. I.R.E.*
MTT-2, p. 100 (Jan., 1954)

53. Wu, T. T. "Theory of the Microstrip", *J. appl. Phys.*
28, 299–302 (March, 1957).

54. Hartwig, C. P. *et al.* "Frequency Dependent Behaviour of Microstrip", *Int. Microwave Symp. (Digest of Papers).*
pp. 110–116, Detroit (1968).

55. Eaves, R. E. and Bolle, D. M. "Guided Waves in Limit Cases of Microstrip", *Trans. I.E.E.E.*
MTT-18, pp. 231–232 (April, 1970).

56. Napoli, L. S. and Hughes, J. J. "High Frequency Behaviour of Microstrip Transmission Lines", *RCA Rev.*
30, 268–276 (June, 1969).

57. Welch, J. D. and Pratt, H. J. "Losses in Microstrip Transmission Systems for Integrated Microwave Circuits", *NEREM Rec.*
8, 100–101, Boston (Nov., 1966).

58. Pucel, R. A. *et al.* "Losses in Microstrip", *Trans. I.E.E.E.*
MTT-16, pp. 342–350 (June, 1968).

59. Hartwig, C. P. *et al.* "Microstrip Technology", *Proc. NEC.*
24, 314–317 (Dec., 1968).

60. Schneider, M. V. "Dielectric Loss in Integrated Microwave Circuits", *B.S.T.J.*
48, 2325–2332 (Sept., 1969).

61. Kitamura, Z. *et al.* "Characteristics of a Stripline with Skin Effect", *Electronics Communs Jap.*
52-A, 34–40 (Sept., 1969).

62. Guckel, H. "Characteristic Impedances of Generalized Rectangular Transmission Lines", *Trans. I.E.E.E.*
MTT-13, pp. 270–274 (May, 1965).

63. Bräckelmann, W. "Wellentypen auf der Streifenleitung mit rechteckigem Schirm", *A.E.Ü.*
21, 641–648 (Dec., 1967).

64. Bräckelmann, W. "Kapazitäten und Induktivitäten gekoppelter Streifenleitungen", *A.E.Ü.*
22, 313–321 (July, 1968).

65. Okean, H. C. "Properties of a TEM Transmission Line Used in Microwave Integrated Circuit Applications", *Trans. I.E.E.E.*
MTT-15, pp. 327–328 (May, 1967).

66. Cohn, S. B. "Slot-line—An alternative transmission medium for integrated circuits", *I.E.E.E. G-MTT Internatl. Microwave Symposium Digest.*
pp. 104–109 (1968).

67. Cohn, S. B. "Slot Line on a Dielectric Substrate", *Trans. I.E.E.E.*
MTT-17, pp. 768–778 (Oct., 1969).

68. Mariani, E. A. *et al.* "Slot Line Characteristics", *Trans. I.E.E.E.*
MTT-17, pp. 1091–1096 (Dec., 1969).

69. Wolfe, P. N. "Capacitance Calculations for Several Simple Two-Dimensional Geometries", *Proc. I.R.E.*
50, 2131–2132 (Oct., 1962).

70. Robinson, G. H. and Allen, J. L. "Slot Line Application to Miniature Ferrite Devices", *Trans. I.E.E.E.*
MTT-17, pp. 1097–1100 (Dec., 1969).

71. Eaves, R. E. and Bolle, D. M. "Modes on Shielded Slot Lines", *A.E.Ü.*
24, 389–394 (Sept., 1970).

72. Wen, C. P. "Coplanar Waveguide: A Surface Strip Transmission Line Suitable for Nonreciprocal Gyromagnetic Device Applications", *Trans. I.E.E.E.*
MTT-17, pp. 1087–1090 (Dec., 1969).

73. Mittra, R. and Itoh, T. "Charge and Potential Distributions in Shielded Striplines", *Trans. I.E.E.E.*
MTT-18, pp. 149–156 (March, 1970).

74. Foster, K. "The Characteristic Impedance and Phase Velocity of High-Q Triplate Line", *J. Br. Instn Radio Engrs.*
18, 715–723 (1958).

75. Cohn, S. B. "Characteristic Impedances of Broadside-Coupled Strip Transmission Lines", *Trans. I.R.E.*
MTT-8, pp. 633–637 (Nov., 1960).

76. Yamamoto, S. *et al.* "Slit-Coupled Strip Transmission Lines", *Trans. I.E.E.E.*
MTT-14, pp. 542–553 (Nov., 1966).

77. Green, H. E. and Pyle, J. R. "The Characteristic Impedance and Velocity Ratio of Dielectric-Supported Strip Line", *Trans. I.E.E.E.*
MTT-13, pp. 135–137 (Jan., 1965).

78. Earle, M. A. and Benedek, P. "Characteristic Impedance of Dielectric Supported Strip Transmission Line", *Trans. I.E.E.E.*
MTT-14, pp. 884–885 (Oct., 1968).

79. Yamashita, E. "Variational Method for the Analysis of Microstrip-Like Transmission Lines", *Trans. I.E.E.E.*
MTT-16, pp. 529–535 (Aug., 1968).

80. Brenner, H. E. "Numerical Solution of TEM-Line Problems Involving Inhomogeneous Media", *Trans. I.E.E.E.* MTT-15, pp. 485–487 (Aug., 1967).

81. Brenner, H. E. "Use a Computer to Design Suspended-Substrate IC's", *Microwaves.* 7, 38–45 (Sept., 1968).

82. Yamashita, E. and Atsuki, K. "Design of Transmission-Line Dimensions for a Given Characteristic Impedance", *Trans. I.E.E.E.* MTT-17, pp. 638–639 (Aug., 1969).

83. Gish, D. L. and Graham, O. "Characteristic Impedance and Phase Velocity of a Dielectric-Supported Air Strip Trans-mission Line with Side Walls", *Trans. I.E.E.E.* MTT-18, pp. 131–148 (March, 1970).

84. Pregla, R. and Schlosser, W. "Waveguide Modes in Dielectrically Supported Strip Lines", *A.E.U.* 22, pp. 379–386 (Aug., 1968).

85. Matthaei, G. L. *et al. Microwave Filters, Impedance-Matching Networks, and Coupling Structures.* McGraw-Hill, New York (1964).

86. Fritsche, H. A. "Wellenwiderstand einer Mikrowellen-Kammerleitung", *Frequenz.* 22, 20–23 (Jan., 1968).

SUMMARY OF USEFUL FORMULAE

Line cross-section	Formulae for $Z_0\sqrt{\kappa}$ (ohms)	Accuracy	Range of validity	Text equation nos.
	$47{\cdot}086\,K'(k)/K(k)$ where $k = \left(\dfrac{p-q}{p+q}\right)^2$ $K(p)/K'(p) = \dfrac{(1-t/b)}{(1+t/b)}$ $q = \sqrt{1-p^2}$	Exact	Unlimited	(3.3.2) (3.3.3) (3.3.5) (3.3.4)
	$\dfrac{47{\cdot}086\,(1-t/b)}{0{\cdot}279+0{\cdot}721t/b}$	1%	$t/b > 0{\cdot}25$	(3.3.6)
	$136{\cdot}7\log_{10}(0{\cdot}9259b/t)$	0·5%	$t/b \leqslant 0{\cdot}5$	(3.3.7)
	$59{\cdot}952\ln\left(\dfrac{1+W'/b}{W/b+t/b}\right)$	10%	$t/b < 0{\cdot}3$ $W/W' < 0{\cdot}8$	(3.4.5)
	$\dfrac{94{\cdot}172}{0{\cdot}558+\{(W/b+t/b)/(1-t/b)\}}$	Not known. Believed better than 20%	Unlimited, providing that $(W'-W) = (b-t)$	(3.4.9)
	$29{\cdot}976\pi\dfrac{K(k)}{K'(k)}$ where $k = \mathrm{sech}\,(\pi W/2b)$	Exact	Unlimited	(3.5.1) (3.5.2)
	$\dfrac{94{\cdot}172}{x(W/b)+(1/\pi)\ln F(x)}$ where $x = \dfrac{1}{1-t/b}$ $F(x) = \dfrac{(x+1)^{x+1}}{(x-1)^{x-1}}$	1%	$\dfrac{W}{b} \geqslant 0{\cdot}35\left(1-\dfrac{t}{b}\right)$ and $\dfrac{t}{b} \leqslant 0{\cdot}25$	(3.5.10) (3.5.12) (3.5.11)

Line cross-section	Formulae for $Z_0\sqrt{\kappa}$ (ohms)	Accuracy	Range of validity	Text equation nos.
	$$\dfrac{188 \cdot 344 \ln X}{\pi + 2\{(W/b - t/b)/(1 - t/b)\} \ln X}$$ where $$X = \dfrac{4}{\pi(t/b)}$$	Better than 1%	$t/b \leqslant 0 \cdot 5$	(3.5.17) (3.5.18)
	$$\dfrac{\kappa}{\kappa_e} 59 \cdot 952 \ln\left(\dfrac{8h}{W} + \dfrac{W}{4h}\right)$$ $$\sqrt{\dfrac{\kappa}{\kappa_e}} \dfrac{119 \cdot 904\pi}{(W/h) + 2 \cdot 42 - 0 \cdot 44(h/W) + (1 - h/W)}$$ where $$\kappa_e = \dfrac{\kappa+1}{2} + \dfrac{\kappa-1}{2}\left(1 + \dfrac{10h}{W}\right)^{-1/2}$$	1% or better 1% or better	$0 \leqslant W/h \leqslant 1$ $1 \leqslant W/h \leqslant 10$	(3.6.9) (3.6.10) (3.6.15)
	$$59 \cdot 952 \sqrt{\dfrac{\kappa}{\kappa_e}} \ln(4h/d)$$ where $\quad d = 0 \cdot 536W + 0 \cdot 67t$ and $\quad \kappa_e = 0 \cdot 475\kappa + 0 \cdot 67$	Approx. 5%	$W/h < 1 \cdot 25$ $0 \cdot 1 < t/W < 0 \cdot 8$ $2 \cdot 5 < \kappa < 6$	(3.6.16) (3.6.18) (3.6.17)
	$$\dfrac{29 \cdot 976\pi(1 - h/b)}{(W/b) + 0 \cdot 4413(1 - (h/b)) - (C/\pi)}$$ where $$C = \left(1 - \dfrac{h}{b}\right)\ln\left(1 - \dfrac{h}{b}\right) + \dfrac{h}{b}\ln\left(\dfrac{h}{b}\right)$$	Better than 1%	$W/h > 0 \cdot 35$ and $\dfrac{W/b}{(1 - h/b)} \geqslant 0 \cdot 35$	(3.8.5) (3.8.6)

4

TRANSMISSION LINES UTILIZING CONDUCTORS OF BOTH CIRCULAR AND RECTANGULAR CROSS-SECTION

4.1 INTRODUCTION

This chapter describes types of lines which may perhaps be regarded as "hybrids" of those discussed in the two preceding chapters, since, of the two conductors which constitute each line, one is circular and the other rectangular in cross-section.

This, unfortunately, increases the difficulties of analysis, and this is reflected in the fact that, to the author's knowledge, no general, exact analysis has yet been given for any of the lines to be described: not even for the most symmetrical form, the square slab-line described in Section 4.2.

Nevertheless, highly accurate formulae and other data are available, and will be presented here.

From a practical constructional viewpoint, the lines treated in this chapter can in certain respects be regarded as combining the advantages of those lines treated in the previous chapters: broadly speaking (depending on tolerances and quantities required) it is easier and cheaper to produce rods of circular cross-section (by turning or grinding processes) than it is to produce rectangular strip conductors (by milling and/or cutting processes). Similarly, it is easier, and usually cheaper, to produce outer conductors having flat, plane bounding surfaces than it is to produce those of cylindrical form, unless the required dimensions of the latter happen to be those of one of the standard production sizes of drawn tubing.

Furthermore, for those applications, such as high-power transmission, which require an appreciable centre conductor thickness (as opposed to those which can make use of printed circuit techniques), a more rugged structure is realizable by the use of rod rather than strip. This is because, for a given Z_0, the lower capacitance engendered by the rounded edges means that a smaller gap is required between the conductors in order to achieve the same capacitance per unit length (and hence Z_0): thus a thicker centre conductor is required.

Lines of the type discussed in this chapter are extensively used in the realization of microwave filters, particularly of the "interdigital" and "comb-line" types [1], and similar structures are used as delay lines in certain types of microwave electron tubes.

The discussion so far has been with reference to lines having circular inner, and rectangular outer, conductors: obviously the "inverse" structure is also feasible, having rectangular inner and circular outer conductors. However, these are seldom encountered in practice, presumably because of the difficulties of accurate construction, and their much lower power-handling capability. The only example which has come to the author's notice, and that only in a theoretical context, is the "coaxial stripline", which is briefly discussed in Section 4.6.

There appears to be no universally-agreed nomenclature for lines with circular inner and rectangular outer conductors, but the term "slab-line" is in fairly wide use, and will be employed here. Note, however, that in at least one instance [2], this term has been applied to lines in which *both* conductors are of *rectangular* cross-section.

4.2   THE SQUARE SLAB-LINE

Following the precedent already established in previous chapters, the most symmetric form of slab-line is treated first, and is illustrated in cross-section in Fig. 4.1. Its rather pleasing symmetry and formal simplicity is deceptive, for although many analyses of this structure have been published over the past thirty years or so, none is both general and exact.

A most interesting comparative discussion of the more well known of these analyses has been given by Cohn [3]: a few of these, together with other treatments, are briefly discussed below.

(a) *Frankel's Analysis* [4]. Frankel appears to have been the first to present an explicit formula for the characteristic impedance of square slab-line: using present notation this is as follows:

$$Z_0\sqrt{\kappa} = 59.952\left\{\ln\left(\frac{4b}{\pi d}\tanh\frac{\pi}{2}\right) - \sum_{m=1}^{\widetilde{}} \ln X_m\right\}\text{ohm} \quad (4.2.1)$$

where

$$X_m = \frac{(\cosh 2m\pi + \cosh \pi)}{(\cosh 2m\pi - \cosh \pi)}\tanh^2 m\pi. \quad (4.2.2)$$

Summing the series, and evaluating the hyperbolic functions, these formulae yield the very

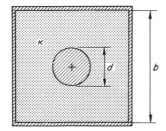

Fig. 4.1. Cross-section of the square slab-line. κ **is the dielectric constant of the medium filling the line interior**

simple expression

$$Z_0\sqrt{\kappa} = 59.952 \ln (1.0787\, b/d)\text{ ohm}. \quad (4.2.3)$$

Equations (4.2.1) and (4.2.2) were obtained by conformal transformation techniques, and strictly speaking relate to a structure consisting of an *almost* circular (elliptical) rod within a square outer conductor. The accuracy of equation (4.2.3) is therefore dependent upon the degree of non-circularity, and hence upon d/b.

According to Cohn [3], the formula is virtually exact for

$$d/b < 0.65 \quad (Z_0 > 30 \text{ ohm})$$

and gives an error within 1.5 per cent for

$$d/b < 0.80$$

Frankel also considers several other structures consisting of rods and parallel planes, and gives formulae for their characteristic impedances: see Section 4.4(b), for example.

(b) *Schelkunoff's Analysis* [5]. Frankel's formulae were obtained by analysis of a "not-quite-circular" conductor within a perfectly square outer conductor. Schelkunoff, on the other hand, considers the case of a perfectly circular conductor within a "not-quite-square" outer conductor; using similar conformal transformation techniques to derive the capacitance per unit length. This can be used to derive the following expression for the characteristic impedance

$$Z_0\sqrt{\kappa} = 59.952 \{\ln (b/d) + 0.0704)\text{ ohm}. \quad (4.2.4)$$

This can obviously be rewritten in the same form as equation (4.2.3), as follows

$$Z_0\sqrt{\kappa} = 59.952 \ln (1.0729\, b/d)\text{ ohm}. \quad (4.2.5)$$

It is apparent that the two results are virtually identical; and both are valid for the same range of d/b.

(c) *Cristal's Analysis* [6, 7]. Using numerical techniques to solve the integral equation associated with the boundary value problem for the rectangular slab-line shown in Fig. 4.2, Cristal has obtained tables and graphs of impedance data, and for the present case of interest, $W' = b$ (which relates to Fig. 4.1) Cohn [3] considers that the data given are highly accurate. Cristal claims an accuracy between 1 and 2 per cent, whereas Carson [19] believes this claim to be unduly pessimistic.

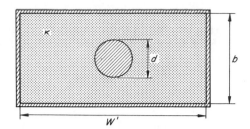

Fig. 4.2. Cross-section of the rectangular slab-line. κ **is the dielectric constant of the medium filling the line interior**

(d) *Lin and Chung's Analysis* [8]. The extensive analyses of transmission lines carried out by these authors have already been referred to in Chapter 3 in connection with the rectangular coaxial line, and they have also given attention to the slab-line structure at present under discussion.

Using conformal transformation techniques, Lin and Chung have derived analytic expressions for upper and lower bounds on the characteristic impedance of square slab-line, which can be written as follows:

$$\text{Upper bound: } Z_0\sqrt{\kappa} = 59\cdot952 \ln U \text{ ohm} \quad (4.2.6a)$$

$$\text{Lower bound: } Z_0\sqrt{\kappa} = 59\cdot952 \ln L \text{ ohm} \quad (4.2.6b)$$

where

$$U^2 = \frac{1 + \text{cn}^2\{(d/\sqrt{2b})\text{K}(0\cdot707107), 1/\sqrt{2}\}}{1 - \text{cn}^2\{(d/\sqrt{2b})\text{K}(0\cdot707107), 1/\sqrt{2}\}} \quad (4.2.7)$$

and

$$L^2 = \frac{1 + \text{cn}\{(d/b)\text{K}(0\cdot707107), 1/\sqrt{2}\}}{1 - \text{cn}\{(d/b)\text{K}(0\cdot707107), 1/\sqrt{2}\}} \quad (4.2.8)$$

In these expressions, "cn" is a Jacobian elliptic function and K is the complete elliptic integral of the 1st kind (see Section 3.2).

Although rather difficult to evaluate without the aid of an electronic computer, equations (4.2.6)–(4.2.8) have two advantages: they are closed-form expressions, and they are exact. Even more important, it has been discovered, quite empirically, that if the geometric mean of equations (4.2.6(a)) and (4.2.6(b)) is taken, the values of $Z_0\sqrt{\kappa}$ thus obtained are (within the appropriate ranges of d/b) in close agreement with those calculated from Frankel's formula, and those obtained numerically by Cristal. Making use of this discovery (and the 4–70 computer!) the data listed in the second column of Table 4.1 have been

calculated (i.e. the case $W'/b = 1$, as illustrated in Fig. 4.1). It is believed that these are the most accurate results yet available. A graphical presentation is given on Fig. 4.3.

It may be mentioned in passing that upper and lower bounds on $Z_0\sqrt{\kappa}$ can also be derived from the work of Anderson and Arthurs [20]: but since these are less restricting than those given by Lin and Chung, and since the geometric mean technique yields no useful results, the work is merely of academic interest in the present context.

Details of other analyses of square slab-line can be obtained by the interested reader from the references listed in [3]: see also [9], [17], and [18].

Conclusion

For practical use, the values listed in Table 4.1 can be regarded as virtually exact.

4.3 THE RECTANGULAR SLAB-LINE

This is a straightforward generalization of the symmetrical structure discussed above, and is illustrated in cross-section in Fig. 4.2.

The reduction in symmetry does, of course, lead to an increase in the difficulties of analysis; and to the author's knowledge, no closed-form solutions, and few numerical solutions, are available. Of the latter, the best known, and most reliable, is probably that of Cristal [6, 7], already mentioned in Section 4.2.

However, Lin and Chung [8] have also applied conformal transformation techniques to the analysis of the rectangular slab-line: the results quoted in Section 4.2 for the square slab-line were simply special cases of this much more general analysis, which has yielded closed algebraic formulae for the upper and lower bounds on the characteristic impedance of rectangular slab-line.

The somewhat arbitrary procedure of taking the geometric mean of the bounding values of $Z_0\sqrt{\kappa}$ has been shown to yield results which are in virtually exact agreement with Cristal's values, as far as can be judged from the graphical presentation given in [7]. Since the accuracy of the latter has been verified by practical application on numerous occasions (see, for example, [10]) it would appear that Lin and Chung's formulae may be used with confidence; though any extensive application of these undoubtedly requires the use of computer facilities. The formulae are as

TABLE 4.1 THE CHARACTERISTIC IMPEDANCES OF SQUARE AND OF RECTANGULAR SLAB-LINE, AS FUNCTIONS OF THE LINE PARAMETERS (THE LATTER ARE DEFINED ON FIG. 4.2). NOTE THAT FURTHER RESULTS, $W^1/b = 5$, ARE GIVEN IN TABLE 4.3 (LAST COLUMN)

$W'/b =$	1·0	1·25	1·50	1·75	2·00	2·50	3·00
d/b	$Z_0\sqrt{\kappa}$ (ohm)	$Z_0\sqrt{\kappa}$ (ohm)	$Z_0\sqrt{\kappa}$ (ohm)	$Z_0\sqrt{\kappa}$ (ohm)	$Z_0\sqrt{\kappa}$ (ohm)	$Z_0\sqrt{\kappa}$ (ohm)	$Z_0\sqrt{\kappa}$ (ohm)
0·05	184·14	189·45	191·95	193·10	193·64	193·99	194·06
0·06	173·21	178·52	181·02	182·17	182·70	183·06	183·13
0·07	163·97	169·28	171·78	172·93	173·46	173·82	173·89
0·08	155·96	161·27	163·77	164·93	165·46	165·81	165·89
0·09	148·90	154·21	156·71	157·87	158·40	158·75	158·82
0·10	142·59	147·89	150·39	151·55	152·08	152·43	152·51
0·11	136·87	142·18	144·68	145·83	146·37	146·72	146·79
0·12	131·66	136·96	139·46	140·62	141·15	141·50	141·58
0·13	126·86	132·16	134·66	135·82	136·35	136·70	136·78
0·14	122·41	127·72	130·22	131·38	131·91	132·26	132·33
0·15	118·28	123·58	126·08	127·24	127·77	128·12	128·20
0·16	114·41	119·71	122·21	123·37	123·90	124·25	124·33
0·17	110·77	116·08	118·58	119·73	120·27	120·62	120·69
0·18	107·35	112·65	115·15	116·31	116·84	117·19	117·27
0·19	104·11	109·41	111·91	113·06	113·60	113·95	114·02
0·20	101·03	106·33	108·83	109·99	110·52	110·87	110·95
0·21	98·11	103·41	105·91	107·06	107·59	107·95	108·02
0·22	95·32	100·62	103·12	104·28	104·80	105·16	105·23
0·23	92·65	97·95	100·45	101·61	102·14	102·49	102·57
0·24	90·10	95·40	97·90	99·05	99·59	99·94	100·01
0·25	87·65	92·95	95·45	96·60	97·14	97·49	97·56
0·26	85·30	90·59	93·09	94·25	94·78	95·14	95·21
0·27	83·04	88·33	90·83	91·99	92·52	92·87	92·95
0·28	80·86	86·15	88·65	89·80	90·34	90·69	90·76
0·29	78·76	84·04	86·54	87·70	88·23	88·58	88·66
0·30	76·72	82·00	84·50	85·66	86·19	86·55	86·62
0·31	74·76	80·03	82·54	83·69	84·23	84·58	84·65
0·32	72·85	78·13	80·63	81·79	82·32	82·67	82·75
0·33	71·01	76·28	78·78	79·94	80·47	80·82	80·90
0·34	69·22	74·48	76·98	78·14	78·68	79·03	79·10

$W'/b =$	1·0	1·25	1·50	1·75	2·00	2·50	3·00
d/b	$Z_0\sqrt{\kappa}$ (ohm)	$Z_0\sqrt{\kappa}$ (ohm)	$Z_0\sqrt{\kappa}$ (ohm)	$Z_0\sqrt{\kappa}$ (ohm)	$Z_0\sqrt{\kappa}$ (ohm)	$Z_0\sqrt{\kappa}$ (ohm)	$Z_0\sqrt{\kappa}$ (ohm)
0·40	59·48	64·70	67·20	68·36	68·89	69·25	69·32
0·41	57·99	63·21	65·71	66·87	67·41	67·76	67·83
0·42	56·55	61·75	64·26	65·42	65·95	66·31	66·38
0·43	55·14	60·33	62·84	64·00	64·53	64·89	64·96
0·44	53·76	58·94	61·45	62·61	63·14	63·50	63·57
0·35	67·48	72·74	75·24	76·40	76·93	77·29	77·36
0·36	65·79	71·04	73·55	74·71	75·24	75·59	75·67
0·37	64·15	69·39	71·90	73·06	73·59	73·94	74·02
0·38	62·55	67·79	70·29	71·45	71·98	72·34	72·41
0·39	60·99	66·22	68·73	69·89	70·42	70·77	70·85
0·45	52·41	57·58	60·09	61·25	61·78	62·14	62·21
0·46	51·10	56·25	58·76	59·92	60·45	60·81	60·88
0·47	49·81	54·95	57·46	58·62	59·15	59·51	59·58
0·48	48·54	53·67	56·18	57·34	57·87	58·23	58·30
0·49	47·31	52·42	54·93	56·09	56·62	56·98	57·05
0·50	46·10	51·19	53·70	54·86	55·39	55·75	55·82
0·51	44·91	49·99	52·50	53·66	54·19	54·54	54·62
0·52	43·74	48·80	51·31	52·47	53·01	53·36	53·43
0·53	42·60	47·64	50·15	51·31	51·84	52·20	52·27
0·54	41·48	46·50	49·01	50·17	50·70	51·05	51·13
0·55	40·38	45·37	47·89	49·04	49·58	49·93	50·00
0·56	39·30	44·27	46·78	47·94	48·47	48·82	48·90
0·57	38·24	43·18	45·69	46·85	47·38	47·73	47·81
0·58	37·19	42·11	44·62	45·78	46·31	46·66	46·73
0·59	36·17	41·06	43·57	44·72	45·25	45·60	45·68
0·60	35·16	40·02	42·53	43·68	44·21	44·56	44·64
0·61	34·17	39·00	41·50	42·66	43·18	43·54	43·61
0·62	33·19	37·98	40·49	41·64	42·17	42·52	42·59
0·63	32·23	36·99	39·49	40·64	41·17	41·52	41·59
0·64	31·29	36·00	38·51	39·66	40·18	40·53	40·60

TABLE 4.1 (continued)

$W'/b =$ 1.0	1.25	1.50	1.75	2.00	2.50	3.00
d/b $Z_0\sqrt{\kappa}$ (ohm)	$Z_0\sqrt{\kappa}$ (ohm)	$Z_0\sqrt{\kappa}$ (ohm)	$Z_0\sqrt{\kappa}$ (ohm)	$Z_0\sqrt{\kappa}$ (ohm)	$Z_0\sqrt{\kappa}$ (ohm)	$Z_0\sqrt{\kappa}$ (ohm)
0.65 30.35	35.03	37.53	38.68	39.21	39.55	39.63
0.66 29.44	34.07	36.57	37.72	38.24	38.59	38.66
0.67 28.53	33.12	35.62	36.76	37.29	37.63	37.71
0.68 27.64	32.19	34.68	35.82	36.34	36.69	36.76
0.69 26.76	31.26	33.75	34.88	35.40	35.75	35.82
0.70 25.90	30.34	32.82	33.96	34.48	34.82	34.89
0.71 25.04	29.43	31.91	33.04	33.56	33.90	33.97
0.72 24.20	28.53	31.00	32.13	32.64	32.99	33.06
0.73 23.37	27.64	30.11	31.23	31.74	32.08	32.15
0.74 22.55	26.76	29.21	30.33	30.84	31.18	31.25
0.75 21.73	25.88	28.33	29.44	29.94	30.28	30.35
0.76 20.93	25.01	27.44	28.55	29.05	29.39	29.46
0.77 20.14	24.15	26.57	27.67	28.17	28.50	28.57
0.78 19.35	23.29	25.70	26.79	27.28	27.61	27.68
0.79 18.58	22.43	24.83	25.91	26.40	26.73	26.79
0.80 17.81	21.58	23.96	25.03	25.52	25.84	25.91
0.81 17.05	20.73	23.09	24.15	24.64	24.96	25.02
0.82 16.29	19.89	22.23	23.28	23.75	24.07	24.13
0.83 15.55	19.05	21.36	22.40	22.87	23.18	23.24
0.84 14.80	18.20	20.49	21.51	21.98	22.28	22.35
0.85 14.07	17.36	19.62	20.63	21.08	21.38	21.45
0.86 13.33	16.52	18.74	19.73	20.18	20.47	20.54
0.87 12.60	15.67	17.86	18.83	19.27	19.56	19.62
0.88 11.87	14.82	16.96	17.91	18.34	18.62	18.68
0.89 11.14	13.96	16.06	16.98	17.40	17.67	17.73
0.90 10.41	13.10	15.14	16.03	16.44	16.71	16.76
0.91 9.68	12.22	14.19	15.06	15.45	15.71	15.77
0.92 8.94	11.33	13.23	14.06	14.44	14.69	14.74
0.93 8.19	10.42	12.23	13.02	13.38	13.62	13.67
0.94 7.42	9.47	11.19	11.94	12.28	12.50	12.55
0.95 6.63	8.49	10.10	10.80	11.11	11.32	11.36
0.96 5.80	7.46	8.93	9.56	9.85	10.04	10.08
0.97 4.91	6.33	7.64	8.20	8.45	8.62	8.65
0.98 3.92	5.07	6.16	6.63	6.84	6.98	7.01
0.99 2.70	3.51	4.31	4.64	4.80	4.90	4.91

follows:

Upper bound: $Z_0\sqrt{\kappa} = 59.952 \ln \dfrac{1}{|Z|_{min}}$ ohm   (4.3.1)

Lower bound: $Z_0\sqrt{\kappa} = 59.952 \ln \dfrac{1}{|Z|_{max}}$ ohm   (4.3.2)

where

$$|Z|^2 = \frac{1 - \mathrm{cn}\,(2\alpha K \cos\theta, k)\,\mathrm{cn}\,(2\alpha K \sin\theta, k')}{1 + \mathrm{cn}\,(2\alpha K \cos\theta, k)\,\mathrm{cn}\,(2\alpha K \sin\theta, k')} \quad (4.3.3)$$

$$\alpha = d/2b, \qquad K = K(k) \quad (4.3.4)$$

and

$$\frac{K'(k)}{K(k)} = \frac{W'}{b} \quad (4.3.5)$$

(see Section 2.2).

The parameter θ is obtained as the solution of the following transcendental equation:

$$\tan\theta = \frac{\mathrm{cn}\,(2\alpha K \cos\theta, k)\,\mathrm{sn}\,(2\alpha K \sin\theta, k')\,\mathrm{dn}\,(2\alpha K \sin\theta, k')}{\mathrm{cn}\,(2\alpha K \sin\theta, k')\,\mathrm{sn}\,(2\alpha K \cos\theta, k)\,\mathrm{dn}\,(2\alpha K \cos\theta, k)}$$

(4.3.6)

where cn and sn are Jacobian Elliptic Functions.

The entire procedure has been programmed for computer evaluation. Results obtained by taking the geometric mean of expressions (4.3.1) and (4.3.2) are tabulated in Table 4.1, for a range of values of W'/b and d/b. A graphical display of the most useful of these results is given in Fig. 4.3.

Conclusion

For practical use, the values listed in Table 4.1

(a)

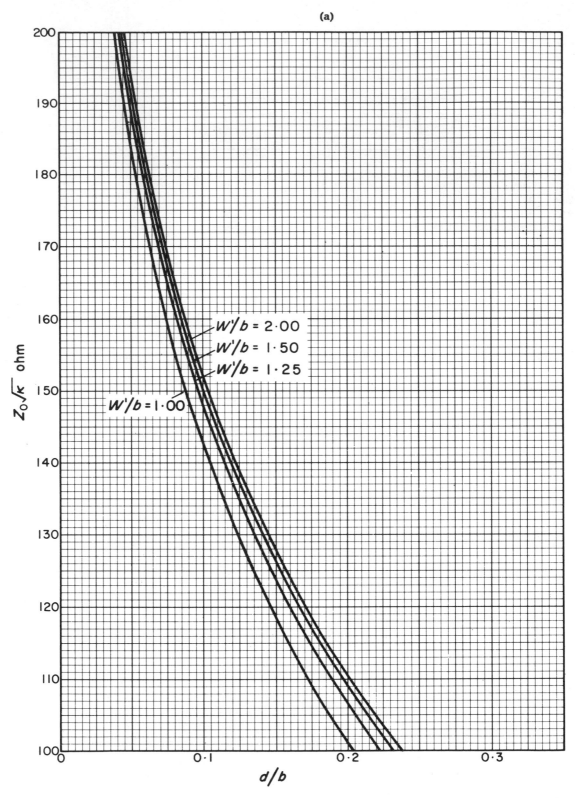

Figs. 4.3(a) and 4.3(b). The characteristic impedance of rectangular slab-
line, as calculated from Lin and Chung's formulae (see text). For definition
of parameters see Fig. 4.2

(b)

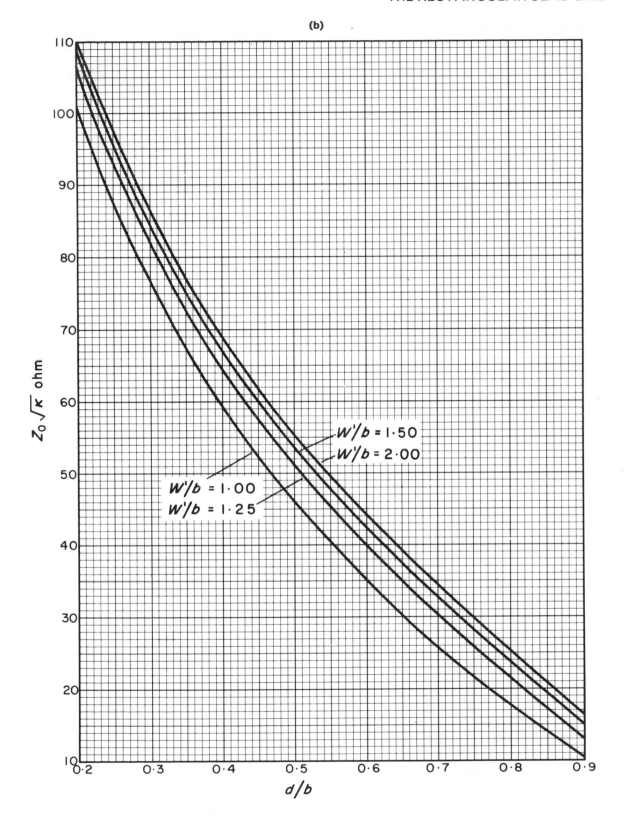

d/b

and graphed on Fig. 4.3 can be regarded as virtually exact.

4.4 THE UNSCREENED SLAB-LINE

This version of slab-line is derived from the rectangular version of Fig. 4.2 by allowing the dimension W' to become infinite: the resulting cross-section is illustrated in Fig. 4.4.

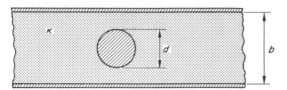

Fig. 4.4. Cross-section of the unscreened slab-line. κ **is the dielectric constant of the medium filling the line interior**

Apart from finding extensive application in microwave filter technology [1, 10], this transmission line has been used in microwave measurement techniques, both as an easily-accessible structure for the detection and measurement of standing waves [11], and also as an accurate primary standard of characteristic impedance [12]. Its analysis has therefore attracted considerable attention, and some of the more important contributions in this area are briefly reviewed below.

(a) *Knight's Analysis* [13]. Knight has derived an expression for the capacitance of a cylinder symmetrically located between two infinite conducting planes. The evaluation of this expression involves some extremely tedious computation and summation of series; however, a few tabulated values are listed in [13], and using these to calculate the characteristic impedance of the structure yields results which are identical with the corresponding values obtained from Wheeler's much simpler formulae [12] (see paragraph (d) below). Knight's result is, using present notation and MKS units,

$$Z_0\sqrt{\kappa} = \frac{59.952}{F(d/b)} \text{ ohm} \qquad (4.4.1)$$

where $F(d/b)$ is tabulated in Table 4.2.

TABLE 4.2   KNIGHT'S CAPACITANCE FUNCTION
$F(d/b)$

d/b	0·1	0·2	0·3	0·4	0·5	0·6	0·7	0·8
$F(d/b)$	0·3931	0·5403	0·6921	0·8653	1·0761	1·3498	1·7369	2·3656

(b) *Frankel's Analysis* [4]. This has already been discussed in Section 4.2 in connection with square slab-line. For the present case of unscreened slab-line, Frankel obtains (by use of conformal transformation techniques) the following simple formula for the characteristic impedance:

$$Z_0\sqrt{\kappa} = 59.952 \ln (4b/\pi d) \text{ ohm} \qquad (4.4.2)$$

The formula is exact for $d \ll b$: an idea of its practical range of validity can be gained from the results in the comparison table, Table 4.3, which lists data calculated from equation (4.4.2) for the entire range of d/b.

(c) *Wholey and Eldred's Analysis* [11]. These authors have utilized conformal transformations to derive an approximate, though quite accurate, formula for the characteristic impedance of unscreened slab-line, which is

$$Z_0\sqrt{\kappa} = 59.952 \{X \ln (\cot \pi d/4b) + XY \ln (\coth \pi d/4b)\} \text{ ohm}$$
$$(4.4.3)$$

where

$$X = \frac{1}{1+Y}$$

and

$$Y = \left\{\frac{\tan^4(\pi d/4b) - 1}{\tanh^4(\pi d/4b) - 1}\right\} \frac{\tanh^2(\pi d/4b)}{\tan^2(\pi d/4b)} \qquad (4.4.4)$$

It will be noted that for small d/b, this rather complicated formula reduces to the simple form quoted by Frankel (equation 4.4.2).

(d) *Wheeler's Analysis* [12]. The results of Wheeler's analysis yield a formula which is probably the most accurate yet proposed (claimed accuracy is 2 parts in 10^7), and at the same time is of comparatively simple form, so that numerical values can be readily obtained. The formula is

$$Z_0\sqrt{\kappa} = 59.952\left\{\ln \frac{(\sqrt{X}+\sqrt{Y})}{\sqrt{X-Y}} - \frac{R^4}{30} + 0.014R^8\right\} \text{ ohm} \qquad (4.4.5)$$

where $X = 1 + 2\sinh^2 R$, $Y = 1 - 2\sin^2 R$, and $R = (\pi/4) \cdot (d/b)$.

Values computed from equation (4.4.5) are tabulated in Table 4.3, together with results from other authors' analyses for comparison. A graphical display is given in Fig. 4.5.

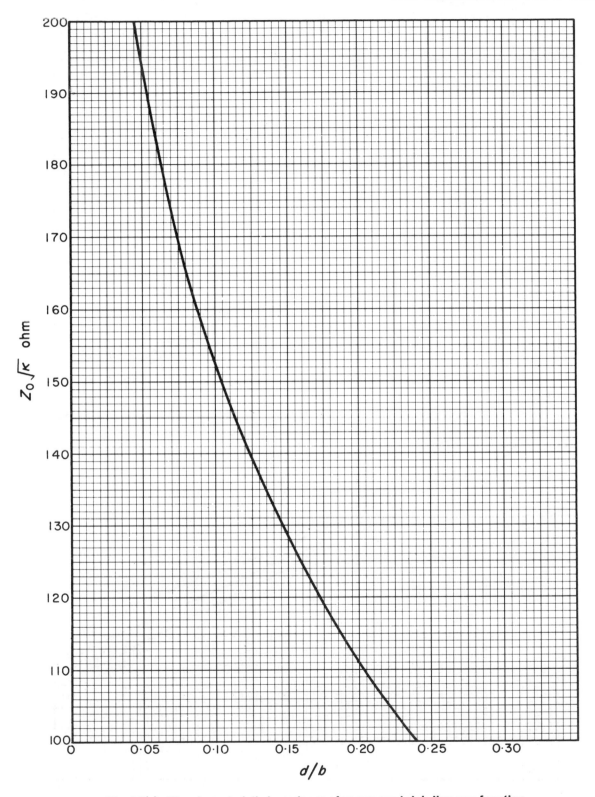

Fig. 4.5(a). The characteristic impedance of unscreened slab-line as a function of the line parameters

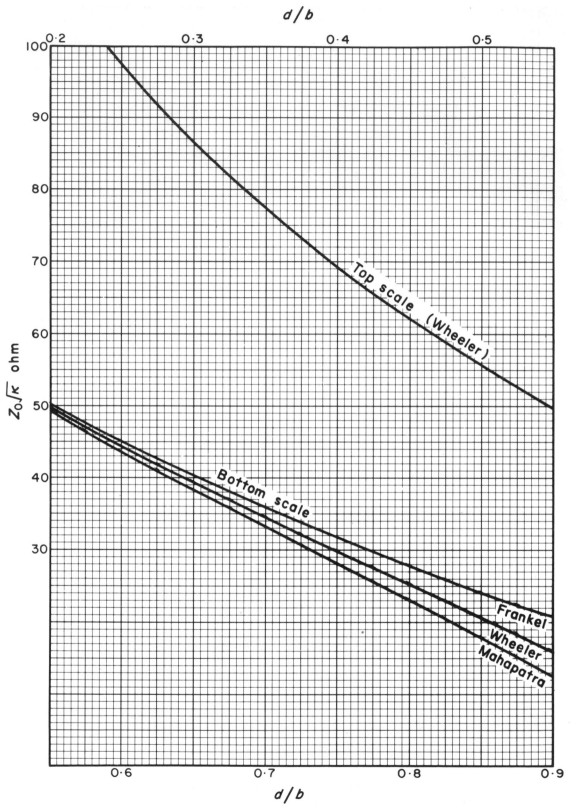

Fig. 4.5(b). The characteristic impedance of unscreened slab-line as a function of the line parameters

(e) *Chisholm's Analysis* [14]. A variational method was used to obtain an expression for the characteristic impedance of "trough line" (see Section 4.5), which consists of a circular cylinder located inside and parallel to the walls of a semi-infinite rectangular trough. Slab-line is a special case of this in which the bottom of the trough is removed to infinity. The analysis is very general, in that the cylinder need not be symmetrically located between the three walls of the trough, and hence the resulting formulae are extremely complex. However, for the case of symmetrical slab-line, considerable simplification is effected, and the resultant formula for the characteristic impedance is

$$Z_0\sqrt{\kappa} = 59{\cdot}952\left\{\ln\left(\frac{4b}{\pi d}\right) - \frac{0{\cdot}01346(d/b)^4}{1 - 0{\cdot}355125(d/b)^4}\right\}\text{ohm.} \quad (4.4.6)$$

It will be noted that the first term in this expression is the same as that derived by Frankel, and that the second term approaches zero for small d/b.

(f) *Mahapatra's Analysis* [15]. Mahapatra uses a conformal mapping to transform the slab-line cross-section to that of a coaxial line with an elliptical centre conductor. He has previously given [16] an approximate expression for the characteristic impedance of the latter structure, and hence derives a formula applicable to unshielded slab-line, which is in some respects reminiscent of Wheeler's formula (equation 4.4.5)

$$Z_0\sqrt{\kappa} = 59{\cdot}952 \ln\left\{\frac{1 + \sqrt{1 - (\tan^2 R - \tanh^2 R)}}{\tan R + \tanh R}\right\}\text{ohm} \quad (4.4.7)$$

where

$$R = \frac{\pi d}{4b}$$

It is claimed that the formula is virtually exact for $d/b < 0{\cdot}6$, but that the error increases rapidly for $d/b > 0{\cdot}75$. This is confirmed by the results tabulated in Table 4.3 (column 6) which were calculated from equation (4.4.7).

(g) *Lin and Chung's Analysis* [8]. These authors' treatment of rectangular slab-line has already been described in Section 4.2, paragraph (d), and it will be apparent from Fig. 4.2 that if the side walls are removed, effectively to infinity, the unshielded structure at present under discussion

is obtained. Hence, for large values of W'/b, Lin and Chung's analysis should yield quite accurate results for unshielded slab-line. That this is indeed so can be seen from the results tabulated in the final column of Table 4.3, which were evaluated by taking the geometric mean of expressions (4.3.1) and (4.3.2), using $W'/b = 5$. That this value of W'/b is effectively "infinite", as required, can be seen by comparing the tabulated results with those for $W'/b = 3$, given in Table 4.1.

Conclusion

For practical use, Wheeler's simple formula (equation (4.4.5)) yields results which are virtually exact: these are tabulated in the second column of Table 4.3.

4.5 THE "TROUGH" LINE

This rather unusual configuration of circular and rectangular conductors (see Fig. 4.6) has not excited as much interest as the structures previously described in this chapter, but it is of some practical importance for the construction of slotted-line devices used in the measurement of voltage standing waves [14, 15] and is also of theoretical interest because its characteristic impedance is precisely half of the odd-mode characteristic impedance of a pair of coupled unshielded slab-lines, as will be seen in Chapter 6.

Although he himself did not specifically do so, Frankel's formulae [4] for parallel wires in rectangular troughs can be used to derive what should probably be regarded as the earliest known formula for the characteristic impedance of trough line.

Frankel derived an expression for the characteristic impedance of two "balanced" (i.e. odd-mode excited: opposite polarities with respect to ground) wires symmetrically located within a rectangular "box". Under such excitation it is obvious that an electric (i.e. perfectly conducting) wall can be introduced midway between the two wires, normal to a line joining their centres, without in any way affecting the existing electric field configuration. If the side walls of the box are now removed to infinity, we obtain in effect either a trough line plus its mirror image, or the two coupled unshielded slab-lines already referred to. In either case, halving the relevant value of characteristic impedance will yield the equivalent

TABLE 4.3 COMPARISON TABLE, GIVING VALUES OF UNSCREENED SLAB-LINE CHARACTERISTIC IMPEDANCE CALCULATED FROM THE VARIOUS FORMULAE (DISCUSSED IN THE TEXT) ORIGINATED BY THE AUTHORS LISTED. THOSE GIVEN BY WHEELER, AND KNIGHT, MAY BE REGARDED AS VIRTUALLY EXACT

d/b	Wheeler $Z_0\sqrt{\kappa}$ (ohm)	Knight $Z_0\sqrt{\kappa}$ (ohm)	Frankel $Z_0\sqrt{\kappa}$ (ohm)	Chisholm $Z_0\sqrt{\kappa}$ (ohm)	Mahapatra $Z_0\sqrt{\kappa}$ (ohm)	Lin and Chung $Z_0\sqrt{\kappa}$ (ohm)
0·10	152·54	152·51	152·53	152·53	152·53	152·53
0·11	146·83		146·81	146·81	146·81	146·81
0·12	141·61		141·60	141·60	141·60	141·60
0·13	136·81		136·80	136·80	136·79	136·80
0·14	132·37		132·35	132·35	132·35	132·35
0·15	128·23		128·22	128·22	128·21	128·22
0·16	124·36		124·35	124·35	124·34	124·35
0·17	120·72		120·71	120·71	120·71	120·71
0·18	117·30		117·29	117·29	117·28	117·29
0·19	114·05		114·05	114·05	114·03	114·04
0·20	110·98	110·96	110·97	110·97	110·95	110·97
0·21	108·05		108·05	108·04	108·03	108·04
0·22	105·26		105·26	105·26	105·23	105·25
0·23	102·59		102·59	102·59	102·56	102·59
0·24	100·03		100·04	100·04	100·01	100·03
0·25	97·58		97·59	97·59	97·55	97·58
0·26	95·23		95·24	95·24	95·19	95·23
0·27	92·96		92·98	92·98	92·92	92·97
0·28	90·78		90·80	90·79	90·73	90·78
0·29	88·67		88·70	88·69	88·62	88·68
0·30	86·63	86·62	86·66	86·65	86·58	86·64
0·31	84·66		84·70	84·69	84·60	84·67
0·32	82·75		82·79	82·79	82·68	82·77
0·33	80·90		80·95	80·94	80·82	80·92
0·34	79·10		79·16	79·15	79·02	79·12
0·35	77·35		77·42	77·41	77·26	77·38
0·36	75·65		75·73	75·72	75·55	75·69
0·37	74·00		74·09	74·07	73·89	74·04
0·38	72·39		72·49	72·47	72·27	72·43
0·39	70·82		70·93	70·91	70·69	70·87
0·40	69·29	69·28	69·42	69·39	69·14	69·34
0·41	67·80		67·94	67·91	67·63	67·85
0·42	66·34		66·49	66·47	66·16	66·40
0·43	64·91		65·08	65·05	64·71	64·98
0·44	63·52		63·70	63·67	63·30	63·59
0·45	62·15		62·35	62·32	61·91	62·23
0·46	60·81		61·04	61·00	60·56	60·90
0·47	59·50		59·75	59·71	59·22	59·60
0·48	58·22		58·49	58·44	57·91	58·32
0·49	56·96		57·25	57·20	56·63	57·07

TABLE 4.3 (continued)

d/b	Wheeler $Z_0\sqrt{\kappa}$ (ohm)	Knight $Z_0\sqrt{\kappa}$ (ohm)	Frankel $Z_0\sqrt{\kappa}$ (ohm)	Chisholm $Z_0\sqrt{\kappa}$ (ohm)	Mahapatra $Z_0\sqrt{\kappa}$ (ohm)	Lin and Chung $Z_0\sqrt{\kappa}$ (ohm)
0·50	55·72	55·71	56·04	55·99	55·36	55·84
0·51	54·50		54·85	54·79	54·12	54·64
0·52	53·31		53·69	53·63	52·89	53·45
0·53	52·14		52·54	52·48	51·69	52·29
0·54	50·98		51·42	51·35	50·50	51·15
0·55	49·85		50·32	50·25	49·33	50·02
0·56	48·73		49·24	49·16	48·17	48·92
0·57	47·63		48·18	48·09	47·03	47·83
0·58	46·54		47·14	47·04	45·90	46·75
0·59	45·48		46·11	46·01	44·79	45·70
0·60	44·42	44·42	45·11	45·00	43·69	44·65
0·61	43·38		44·12	44·00	42·59	43·63
0·62	42·35		43·14	43·02	41·51	42·61
0·63	41·34		42·18	42·05	40·44	41·61
0·64	40·33		41·24	41·09	39·38	40·62
0·65	39·34		40·31	40·15	38·33	39·65
0·66	38·36		39·39	39·23	37·28	38·68
0·67	37·39		38·49	38·32	36·24	37·72
0·68	36·42		37·60	37·42	35·21	36·78
0·69	35·47		36·73	36·53	34·18	35·84
0·70	34·52	34·52	35·87	35·65	33·16	34·91
0·71	34·58		35·02	34·79	32·14	33·99
0·72	32·65		34·18	33·94	31·12	33·08
0·73	31·73		33·35	33·09	30·11	32·17
0·74	30·80		32·53	32·26	29·10	31·27
0·75	29·89		31·73	31·44	28·09	30·37
0·76	28·98		30·94	30·63	27·08	29·47
0·77	28·07		30·15	29·83	26·07	28·58
0·78	27·16		29·38	29·03	25·06	27·70
0·79	26·26		28·61	28·25	24·04	26·81
0·80	25·35	25·34	27·86	27·47	23·03	25·93
0·81	24·45		27·12	26·71	22·01	25·04
0·82	23·54		26·38	25·95	20·98	24·15
0·83	22·63		25·65	25·19	19·95	23·26
0·84	21·72		24·94	24·45	18·91	22·36
0·85	20·80		24·23	23·71	17·86	21·46
0·86	19·87		23·52	22·98	16·81	20·55
0·87	18·93		22·83	22·25	15·74	19·63
0·88	17·98		22·15	21·53	14·66	18·70
0·89	17·01		21·47	20·82	13·57	17·75
0·90	16·03		20·80	20·11	12·46	16·78
0·91	15·02		20·14	19·40	11·33	15·78
0·92	13·98		19·48	18·71	10·19	14·75
0·93	12·90		18·83	18·01	9·02	13·68
0·94	11·77		18·19	17·32	7·83	12·56

TABLE 4.3 (continued)

d/b	Wheeler $Z_0\sqrt{\kappa}$ (ohm)	Knight $Z_0\sqrt{\kappa}$ (ohm)	Frankel $Z_0\sqrt{\kappa}$ (ohm)	Chisholm $Z_0\sqrt{\kappa}$ (ohm)	Mahapatra $Z_0\sqrt{\kappa}$ (ohm)	Lin and Chung $Z_0\sqrt{\kappa}$ (ohm)
0·95	10·58		17·56	16·63	6·62	11·37
0·96	9·29		16·93	15·95	5·37	10·09
0·97	7·87		16·31	15·27	4·09	8·66
0·98	6·24		15·69	14·59	2·77	7·02
0·99	4·17		15·08	13·91	1·41	4·92

TABLE 4.4(a)   VALUES OF NUMERICAL COEFFICIENTS FOR USE IN CHISHOLM'S FORMULA FOR THE CHARACTERISTIC IMPEDANCE OF TROUGH LINE

h/b	j	N_j 1	2	3	D_j 1	2	3
0·25		−0·2966	−0·7312	+8·680	−6·321	−54·90	+174·8
0·50		−0·01177	−0·04932	+0·3353	−2·503	−10·11	+ 21·58
0·75		−0·0005068	−0·1714	+0·3220	−1·822	− 6·559	+ 11·80
∞		0	−0·2153	0	0	− 5·682	0

Note: $j = 1, 2, 3$.

result for single wire trough line.

Making these modifications to Frankel's equation (18) [4] leads to the following result:

$$Z_0\sqrt{\kappa} = 59\cdot952 \ln\left\{\frac{4b}{\pi d}\tanh\left(\frac{\pi h}{b}\right)\right\} \text{ohm.} \quad (4.5.1)$$

This expression has been used by Wheeler [9] in preparing a set of transmission-line impedance curves.

A very general analysis of trough line has been given by Chisholm [14], using a variational method to develop an expression for the characteristic impedance of a structure in which the location of the cylindrical wire conductor is not restricted to a centrally-symmetric position. This is of some value in determining the effects of constructional tolerances, but the relevant formulae are rather complex and unwieldy, and will not be quoted here. Of more general interest is the somewhat simpler expression for the characteristic impedance of the symmetric trough line (see [14]):

$$Z_0\sqrt{\kappa} = 59\cdot952 \ln\left\{\frac{4b}{\pi d}\tanh\left(\frac{\pi h}{b}\right)\right\} + F\left(\frac{d}{b},\frac{h}{b}\right) \text{ohm.} \quad (4.5.2)$$

Comparing this with Frankel's formula (equation (4.5.1)) it can be seen that the variational approach has yielded an extra term, given by Chisholm [14] as

$$F\left(\frac{d}{b},\frac{h}{b}\right) = 376\cdot687R^2\frac{(N_1+N_2R^2+N_3R^4)}{1+D_1R^2+D_2R^4+D_3R^6} \quad (4.5.3)$$

where $R = d/b$ and the N_j and D_j ($j = 1, 2, 3$) are numerical coefficients. These have been evaluated by Chisholm for most cases of practical interest, and his results are listed in Table 4.4(a). Table 4.4(b) compares results calculated from Frankel's and Chisholm's formulae.

TABLE 4.4(b)   COMPARISON TABLE GIVING VALUES OF TROUGH LINE CHARACTERISTIC IMPEDANCE CALCULATED FROM FRANKEL'S AND CHISHOLM'S FORMULAE. FOR DEFINITION OF PARAMETERS SEE FIG. 4.6

d/b	$h/b = 1/4$ Frankel $Z_0\sqrt{\kappa}$	Chisholm	$h/b = 1/2$ Frankel $Z_0\sqrt{\kappa}$	Chisholm	$h/b = 3/4$ Frankel $Z_0\sqrt{\kappa}$	Chisholm	$h/b = \infty$ Frankel $Z_0\sqrt{\kappa}$	Chisholm
0·05	168·79	168·72	188·90	188·89	193·01	193·01	194·08	194·08
0·06	157·86	157·76	177·97	177·96	182·07	182·07	183·15	183·15
0·07	148·62	148·48	168·73	168·72	172·83	172·83	173·91	173·91
0·08	140·61	140·43	160·72	160·71	164·83	164·83	165·90	165·90
0·09	133·55	133·32	153·66	153·65	157·77	157·77	158·84	158·84

TABLE 4.4(b) (continued)

| | $h/b = 1/4$ | | $h/b = 1/2$ | | $h/b = 3/4$ | | $h/b = \infty$ | |
| | Frankel | Chisholm | Frankel | Chisholm | Frankel | Chisholm | Frankel | Chisholm |
d/b	$Z_0\sqrt{\kappa}$		$Z_0\sqrt{\kappa}$		$Z_0\sqrt{\kappa}$		$Z_0\sqrt{\kappa}$	
0·10	127·23	126·95	147·34	147·33	151·45	151·45	152·53	152·53
0·11	121·52	121·17	141·63	141·61	145·74	145·73	146·81	146·81
0·12	116·30	115·89	136·41	136·40	140·52	140·52	141·60	141·60
0·13	111·50	111·01	131·61	131·59	135·72	135·72	136·80	136·80
0·14	107·06	106·49	127·17	127·15	131·28	131·27	132·35	132·35
0·15	102·92	102·26	123·03	123·01	127·14	127·14	128·22	128·22
0·16	99·05	98·30	119·16	119·13	123·27	123·27	124·35	124·35
0·17	95·42	94·56	115·53	115·50	119·64	119·63	120·71	120·71
0·18	91·99	91·02	112·10	112·07	116·21	116·20	117·29	117·28
0·19	88·75	87·66	108·86	108·82	112·97	112·96	114·05	114·04
0·20	85·68	84·45	105·79	105·74	109·89	109·89	110·97	110·96
0·21	82·75	81·39	102·86	102·81	106·97	106·96	108·05	108·04
0·22	79·96	78·45	100·07	100·01	104·18	104·17	105·26	105·25
0·23	77·30	75·62	97·41	97·34	101·52	101·50	102·59	102·58
0·24	74·75	72·90	94·86	94·79	98·96	98·95	100·04	100·02
0·25	72·30	70·27	92·41	92·33	96·52	96·50	97·59	97·57
0·26	69·95	67·73	90·06	89·97	94·16	94·14	95·24	95·22
0·27	67·69	65·26	87·79	87·70	91·90	91·88	92·98	92·95
0·28	65·50	62·85	85·61	85·52	89·72	89·69	90·80	90·77
0·29	63·40	60·51	83·51	83·40	87·62	87·59	88·70	88·66
0·30	61·37	58·22	81·48	81·36	85·59	85·55	86·66	86·62
0·31	59·40	55·99	79·51	79·39	83·62	83·58	84·70	84·65
0·32	57·50	53·79	77·61	77·48	81·72	81·67	82·79	82·74
0·33	55·65	51·63	75·76	75·62	79·87	79·82	80·95	80·89
0·34	53·86	49·51	73·97	73·82	78·08	78·02	79·16	79·09
0·35	52·13	47·41	72·24	72·07	76·34	76·28	77·42	77·34
0·36	50·44	45·33	70·55	70·37	74·66	74·58	75·73	75·65
0·37	48·80	43·27	68·90	68·72	73·01	72·93	74·09	73·99
0·38	47·20	41·21	67·31	67·11	71·41	71·32	72·49	72·38
0·39	45·64	39·16	65·75	65·54	69·86	69·75	70·93	70·82
0·40	44·12	37·11	64·23	64·01	68·34	68·23	69·42	69·28
0·41	42·64	35·05	62·75	62·51	66·86	66·73	67·94	67·79
0·42	41·20	32·96	61·31	61·05	65·41	65·28	66·49	66·33
0·43	39·79	30·85	59·90	59·63	64·00	63·85	65·08	64·90
0·44	38·41	28·70	58·52	58·23	62·62	62·46	63·70	63·51
0·45	37·06	26·49	57·17	56·87	61·28	61·10	62·35	62·14
0·46	35·74	24·22	55·85	55·53	59·96	59·76	61·04	60·81
0·47	34·45	21·86	54·56	54·23	58·67	58·46	59·75	59·50
0·48	33·19	19·39	53·30	52·95	57·41	57·18	58·49	58·21
0·49	31·95	16·77	52·06	51·69	56·17	55·92	57·25	56·95
0·50	30·74		50·85	50·46	54·96	54·69	56·04	55·71
0·51	29·56		49·67	49·25	53·77	53·48	54·85	54·50
0·52	28·39		48·50	48·06	52·61	52·29	53·69	53·31
0·53	27·25		47·36	46·90	51·47	51·12	52·54	52·13
0·54	26·13		46·24	45·76	50·35	49·98	51·42	50·98

TABLE 4.4(b) (continued)

| | h/b = 1/4 | | h/b = 1/2 | | h/b = 3/4 | | h/b = ∞ | |
| | Frankel | Chisholm | Frankel | Chisholm | Frankel | Chisholm | Frankel | Chisholm |
d/b	$Z_0\sqrt{\kappa}$		$Z_0\sqrt{\kappa}$		$Z_0\sqrt{\kappa}$		$Z_0\sqrt{\kappa}$	
0·55	25·03		45·14	44·63	49·25	48·85	50·32	49·84
0·56	23·95		44·06	43·53	48·17	47·74	49·24	48·73
0·57	22·89		43·00	42·44	47·11	46·64	48·18	47·63
0·58	21·85		41·95	41·37	46·06	45·57	47·14	46·54
0·59	20·82		40·93	40·31	45·04	44·51	46·11	45·47
0·60	19·81		39·92	39·28	44·03	43·46	45·11	44·42
0·61	18·82		38·93	38·26	43·04	42·43	44·12	43·38
0·62	17·85		37·96	37·25	42·06	41·41	43·14	42·35
0·63	16·89		37·00	36·26	41·11	40·41	42·18	41·34
0·64	15·94		36·05	35·28	40·16	39·41	41·24	40·33
0·65	15·01		35·12	34·31	39·23	38·43	40·31	39·34
0·66	14·10		34·21	33·36	38·32	37·46	39·39	38·36
0·67	13·20		33·31	32·42	37·41	36·51	38·49	37·39
0·68	12·31		32·42	31·49	36·53	35·56	37·60	36·43
0·69	11·43		31·54	30·57	35·65	34·62	36·73	35·48
0·70	10·57		30·68	29·67	34·79	33·69	35·87	34·54
0·71	9·72		29·83	28·77	33·94	32·77	35·02	33·60
0·72	8·88		28·99	27·89	33·10	31·86	34·18	32·67
0·73	8·06		28·17	27·01	32·27	30·95	33·35	31·75
0·74	7·24		27·35	26·14	31·46	30·05	32·53	30·83
0·75	6·44		26·54	25·29	30·65	29·16	31·73	29·92
0·76	5·64		25·75	24·44	29·86	28·27	30·94	29·02
0·77	4·86		24·97	23·60	29·07	27·39	30·15	28·12
0·78	4·08		24·19	22·76	28·30	26·51	29·38	27·22
0·79	3·32		23·43	21·94	27·54	25·64	28·61	26·32
0·80	2·57		22·68	21·12	26·78	24·77	27·86	25·43
0·81	1·82		21·93	20·31	26·04	23·90	27·12	24·54
0·82	1·09		21·20	19·50	25·30	23·03	26·38	23·65
0·83	0·36		20·47	18·70	24·58	22·17	25·65	22·76
0·84			19·75	17·91	23·86	21·30	24·94	21·87
0·85			19·04	17·12	23·15	20·44	24·23	20·98
0·86			18·34	16·34	22·45	19·57	23·52	20·08
0·87			17·65	15·56	21·75	18·70	22·83	19·19
0·88			16·96	14·78	21·07	17·83	22·15	18·28
0·89			16·28	14·01	20·39	16·95	21·47	17·38
0·90			15·61	13·24	19·72	16·07	20·80	16·46
0·91			14·95	12·48	19·06	15·18	20·14	15·54
0·92			14·30	11·72	18·40	14·28	19·48	14·61
0·93			13·65	10·95	17·76	13·38	18·83	13·67
0·94			13·01	10·19	17·11	12·46	18·19	12·72

4.6 THE COAXIAL STRIPLINE

As was briefly mentioned in the Introduction to this chapter (Section 4.1), this structure (see Fig. 4.7), although a feasible form of microwave transmission line, has not to the author's knowledge found any practical application, and it is included here mainly for completeness, as it constitutes the "limiting form" of one of the suspended-substrate microstrip lines discussed in Chapter 3 (see Section 3.7.2). This is arrived at by reducing

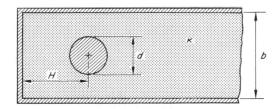

Fig. 4.6. Cross-section of the trough line. κ is the dielectric constant of the medium filling the line interior

the substrate thickness to zero, and hence filling the entire line interior with a single homogeneous dielectric of dielectric constant κ: this yields the structure shown in Fig. 4.7.

A straightforward conformal transformation analysis leads to the following formula for characteristic impedance:

$$Z_0\sqrt{\kappa} = 94 \cdot 172 \frac{K'(x)}{K(x)} \text{ ohm} \qquad (4.6.1)$$

where

$$x = \frac{2(W/b)}{1+(W/b)^2} \qquad (4.6.2)$$

A similar, though not identical, result can be obtained from equations (3.7.1), (3.7.3), and (3.7.4)

by carrying out the limiting process just described, i.e. zero substrate thickness ($h = 0$) and one dielectric ($\kappa' = \kappa$). The formulae thus derived are:

$$Z_0\sqrt{\kappa} = 188 \cdot 344 \frac{K'(k)}{K(k)} \text{ ohm} \qquad (4.6.3)$$

where

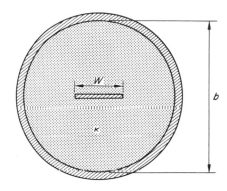

Fig. 4.7. Cross-section of the coaxial stripline. κ is the dielectric constant of the medium filling the line interior

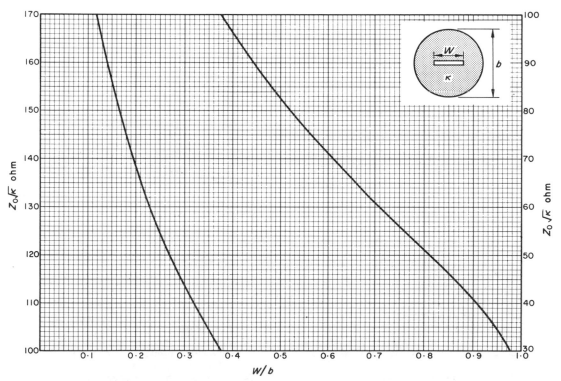

Fig. 4.8. The characteristic impedance of the coaxial stripline as a function of the line parameters. κ is the dielectric constant of the medium filling the line interior

$$(1-k^2)^{1/2} = \frac{(1-W/b)^2}{(1+W/b)^2} \qquad (4.6.4)$$

Although these differ from the two preceding formulae, substitution of numerical data quickly demonstrates that they do in fact yield the same results.

This line has also been treated, as part of a more general analysis, by Boyd [21]. By simple algebraic manipulation, the relevant formulae can be shown to be identical with equations (4.6.1) and (4.6.2).

A graph of $Z_0\sqrt{\kappa}$ versus W/b is given in Fig. 4.8, plotted from the data listed in Table 4.5.

TABLE 4.5   THE CHARACTERISTIC IMPEDANCE OF THE COAXIAL STRIPLINE, AS A FUNCTION OF THE LINE PARAMETERS (THE LATTER ARE DEFINED ON FIG. 4.7)

W/b	$Z_0\sqrt{\kappa}$	W/b	$Z_0\sqrt{\kappa}$	W/b	$Z_0\sqrt{\kappa}$	W/b	$Z_0\sqrt{\kappa}$
0.05	221.15	0.50	82.63	0.25	124.64	0.70	60.95
0.06	210.22	0.51	81.40	0.26	122.28	0.71	59.97
0.07	200.98	0.52	80.20	0.27	120.01	0.72	59.00
0.08	192.98	0.53	79.01	0.28	117.83	0.73	58.02
0.09	185.92	0.54	77.84	0.29	115.71	0.74	57.05
0.10	179.60	0.55	76.69	0.30	113.67	0.75	56.09
0.11	173.88	0.56	75.55	0.31	111.70	0.76	55.12
0.12	168.67	0.57	74.43	0.32	109.79	0.77	54.15
0.13	163.87	0.58	73.33	0.33	107.93	0.78	53.19
0.14	159.42	0.59	72.24	0.34	106.13	0.79	52.22
0.15	155.29	0.60	71.16	0.35	104.38	0.80	51.24
0.16	151.42	0.61	70.09	0.36	102.68	0.81	50.27
0.17	147.78	0.62	69.04	0.37	101.02	0.82	49.28
0.18	144.35	0.63	68.00	0.38	99.41	0.83	48.29
0.19	141.11	0.64	66.97	0.39	97.83	0.84	47.29
0.20	138.03	0.65	65.94	0.40	96.29	0.85	46.27
0.21	135.10	0.66	64.93	0.41	94.79	0.86	45.25
0.22	132.31	0.67	63.93	0.42	93.33	0.87	44.20
0.23	129.64	0.68	62.93	0.43	91.89	0.88	43.14
0.24	127.09	0.69	61.94	0.44	90.49	0.89	42.05
				0.45	89.12	0.90	40.93
				0.46	87.77	0.91	39.77
				0.47	86.45	0.92	38.57
				0.48	85.15	0.93	37.32
				0.49	83.88	0.94	36.00

W is the width of strip and b is the diameter of coaxial line

REFERENCES

1. Matthaei, G. L., Young, L., and Jones, E. M. T. *Microwave Filters, Impedance-Matching Networks, and Coupling Structures.*
   Chs 8 and 10, McGraw-Hill, New York (1964).
2. Bates, R. H. T. "The Characteristic Impedance of the Shielded Slab Line", *Trans. I.R.E.*
   MTT-4, pp. 28–33 (Jan., 1956).
3. Cohn, S. B. "Beating a Problem to Death", *Microwave J.* **12**, 22–24 (Nov., 1969).
4. Frankel, S. "Characteristic Impedance of Parallel Wires in Rectangular Troughs", *Proc. I.R.E.* **30**, 182–190 (April, 1942).
5. Schelkunoff, S. A. *Applied Mathematics for Engineers and Scientists.*
   2nd edn, pp. 298–300, Van Nostrand, Princeton (1965).
6. Cristal, E. G. "Coupled Circular Cylindrical Rods Between

Parallel Ground Planes", *Trans. I.E.E.E.* MTT-12, pp. 428–439 (July, 1964).

7. Cristal, E. G. "Characteristic Impedances of Coaxial Lines of Circular Inner and Rectangular Outer Conductors", *Proc. I.E.E.E.* **52**, 1265–1266 (Oct., 1964).

8. Lin, W-G. and Chung, S-L. "A New Method of Calculating the Characteristic Impedances of Transmission Lines", *Acta phys. sin.* **19**, 249–258 (April, 1963) (in Chinese).

9. Wheeler, H. A. "Transmission-Line Impedance Curves", *Proc. I.R.E.* **38**, 1400–1403 (Dec., 1950).

10. Nicholson, B. F. "The Practical Design of Interdigital and Comb-line Filters", *Radio and Electronic Engr.* **34**, 39–52 (July, 1967).

11. Wholey, W. B. and Eldred, W. N. "A New Type of Slotted Line Section", *Proc. I.R.E.* **38**, 244–248 (March, 1950).

12. Wheeler, H. A. "The Transmission-Line Properties of a Round Wire between Parallel Planes", *Trans. I.R.E.* AP-3, pp. 203–207 (Oct., 1955).

13. Knight, R. C. "The Potential of a Circular Cylinder between Two Infinite Planes", *Proc. math. Soc.* **39**, 272–281 (Dec. 14, 1933).

14. Chisholm, R. M. "The Characteristic Impedance of Trough and Slab Lines", *Trans. I.R.E.* MTT-4, pp. 166–172 (July, 1956).

15. Mahapatra, S. "Characteristic Impedance of a Slab-Line", *Proc. I.R.E.* **48**, 1652–1653 (Sept., 1960).

16. Mahapatra, S. "Coaxial Transmission Lines", *Electronic and Radio Engr.* **35**, 63–67 (Feb., 1958).

17. *Microwave Engineers Technical and Buyers Guide.* pp. 19 and 24, Horizon House, Dedham, Mass. (Feb., 1970).

18. Seshagiri, N. "Least-Weighted-Square Method for Analysis and Synthesis of Transmission Lines", *Trans. I.E.E.E.* MTT-15, p. 497 (Sept., 1967).

19. Carson, C. T. "The Numerical Solution of TEM Mode Transmission Lines with Curved Boundaries", *Trans. I.E.E.E.* MTT-15, pp. 269–70 (April, 1967).

20. Anderson, N. and Arthurs, A. M. "Bounds for Capacities in microwave filter problems", *Int. J. Electron.* **28**, pp. 259–262 (March, 1970).

21. Boyd, C. R. "Characteristic Impedance of Multifin Transmission Lines", *Trans. I.E.E.E.* MTT-15, pp. 487–88 (August, 1967).

SUMMARY OF USEFUL FORMULAE

Structure	Formula for $Z_0\sqrt{\kappa}/59\cdot952$	Range of validity	Text equation no.
	$\ln(1\cdot0787b/d)$	Exact for $\quad d/b < 0\cdot65$ Within 1·5% for $d/b < 0\cdot8$	(4.2.3)
	None known: see Table 4.1		
	$\ln(4b/\pi d)$	Within 1% for $d/b < 0\cdot55$ Within 5% for $d/b < 0\cdot75$	(4.4.2)
	$\ln\left(\dfrac{\sqrt{X}+\sqrt{Y}}{\sqrt{X-Y}}\right)-\dfrac{R^4}{30}+0\cdot014R^8$ where $\quad X = 1+2\sinh^2 R$ $Y = 1-2\sin^2 R$ $R = (\pi d/4b)$	Virtually exact	(4.4.5)
	$\ln\left\{\dfrac{4b}{\pi d}\tanh\left(\dfrac{\pi h}{b}\right)\right\}$	Not known. Probably very accurate for $d/b < 0\cdot5,\ h/b < 0\cdot5$	(4.5.1)
	$\dfrac{\pi}{2}\dfrac{K'(x)}{K(x)}$ where $\quad x = \dfrac{2(W/b)}{1+(W/b)^2}$	Exact for zero thickness strip	(4.6.1) (4.6.2)

5

TRANSMISSION LINES OF "UNUSUAL" CROSS-SECTION

5.1 INTRODUCTION

In a sense, this chapter is a "catalogue of curiosities" consisting as it does of descriptions of transmission lines which have found application only in highly unusual or specialized situations, or which, in some instances, have never actually been used, but have simply been proposed as possibilities.

The main purpose in presenting such information, apart from the natural desire to achieve as complete a coverage of the book's topic as possible, is to bring it to the attention of the development engineer seeking possible solutions to specialized transmission-line problems.

Naturally, the more "unusual" the line cross-section, the more difficult it is to analyze, and for this reason the data and information presented below is, in some instances, rather less comprehensive and/or less accurate than that given in the other chapters of the book. Nevertheless, it is felt that a useful service will have been performed if the reader is simply made aware of the tremendous range of unusual types of transmission lines which have been proposed, and is provided with sufficient information to enable him to carry out more detailed investigations of those which may be of use in a particular application.

In contrast to the other chapters, which are to a large extent intended to serve as sources of information and data for immediate use, it is hoped that this chapter will rather serve as a "source of ideas".

5.2 THE ELLIPTIC COAXIAL LINE

As was stated in Chapter 2 (Section 2.1) the elliptic coaxial transmission line should, from a strictly logical point of view, have been included therein, since it is the most general form of coaxial line from which all the other Chapter 2 types are derived.

However, since its conductors are not of circular cross-section, and since it is in comparatively limited use, it was considered more appropriate to include it in the present chapter.

For the case in which the conductor cross-sections are confocal ellipses, as shown in Fig. 5.1, Smythe [1] gives, as the answer to a set problem, an expression for the inductance per unit length. Using this to derive the formula for characteristic impedance yields

$$Z_0\sqrt{\kappa} = 59{\cdot}952 \ln\left(\frac{W'+b}{W+t}\right)\text{ohm.} \qquad (5.2.1)$$

A complete modal analysis of the structure, using Maxwell's equations in confocal elliptic

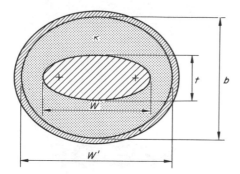

Fig. 5.1. Cross-section of the elliptic coaxial line. The conductor cross-sections are confocal ellipses, with major axes W' and W, and minor axes b and t, respectively. κ is the dielectric constant of the medium filling the line interior

coordinates, has been given by King and Wiltse [3], and yields (after minor approximations involving the use of the asymptotic expressions for the Bessel functions J_n and Y_n) a formula which can be written in a form identical to equation (5.2.1).

Note that as $W' \to b$ and $W \to t$, the expression correctly reduces to that for the simple circular coaxial line (see equation (2.2.2)).

In a further problem Smythe [2] gives a formula for the capacitance per unit length of a structure which can approximately be regarded as an extreme form of the elliptic coaxial line, in which $t = 0$. This formula can be used to derive the following expression for the characteristic impedance of the structure:

$$Z_0\sqrt{\kappa} = 59 \cdot 952 \cosh^{-1}(W'/W) \text{ ohm.} \qquad (5.2.2)$$

A further special case of the elliptic coaxial line is that in which the outer conductor is in fact circular in cross-section. This has been analyzed by Mahapatra [4], who gives the following approximate expression for its characteristic impedance:

$$Z_0\sqrt{\kappa} = 59 \cdot 952 \ln \left\{ \frac{1+\sqrt{1-(W/b)^2+(t/b)^2}}{(W/b+t/b)} \right\} \text{ ohm.} \qquad (5.2.3)$$

In general, values calculated from this expression are rather larger than the "true" figures, but Mahapatra claims an error of less than 1 per cent provided that

$$\left(\frac{W}{b}\right)^2 - \left(\frac{t}{b}\right)^2 \leqslant 0 \cdot 06.$$

The preceding information has all been relevant to TEM mode propagation in elliptic coaxial lines. However, as in the case of the simple circular coaxial line (Chapter 2) at sufficiently high frequencies, higher-order modes of the TE and TM type can propagate. Information on these can be found in [7] and [8]. The latter source in particular gives extensive tabulations of the cut-off wavelengths of many of the TE_{mn} and TM_{mn} modes as functions of t/b and b/W'.

5.3 THE PARTIALLY-FILLED COAXIAL LINE

Two versions of this line are illustrated in cross-section in Fig. 5.2. They are of some value in the design of stepped impedance transformers.

The type shown in Fig. 5.2(a) is mathematically the simpler of the two, and an approximate analysis has been given by Angelakos [5]. Of course, neither type of line can propagate a pure TEM mode, because of the presence of dielectric interfaces at which the boundary conditions cannot be satisfied by such a mode. However, as in previous similar cases (e.g. microstrip) the field configuration can be regarded as "pseudo"-TEM, and an analysis carried out under this assumption.

Angelakos' result for the structure of Fig. 5.2(a) can be written as follows:

$$Z_0\sqrt{\kappa} = \frac{59 \cdot 952 \ln (b/d)}{\sqrt{(\theta/2\pi)\{(\kappa'/\kappa)-1\}+1}} \text{ ohm.} \qquad (5.3.1)$$

The structure shown in Fig. 5.2(b) has been proposed by Sullivan and Parkes [6] as having

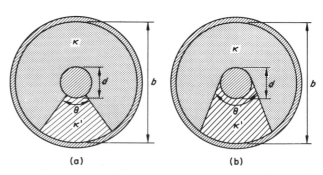

(a)   (b)

Fig. 5.2. Cross-sectional views of two versions of the partially-filled coaxial line. The centre conductor is supported on a segment of dielectric, of angle θ and dielectric constant κ'. The remainder of the line interior is filled with a dielectric medium, usually air, of dielectric constant κ

the practical advantages of easier and stronger construction. An approximate analysis [6], based upon the derivation of the total electrostatic capacitance of the structure, yields the following expression for the characteristic impedance of the "pseudo"-TEM mode:

$$Z_0\sqrt{\kappa} = \frac{59\cdot952 \ln (b/d)}{\sqrt{\theta/2\pi[\{\kappa' \ln (b/d)\}/\{\kappa \ln (c/d)+\kappa' \ln (b/c)\}-1]+1}}$$

$$\text{ohm} \quad (5.3.2)$$

where κ and κ' are the dielectric constants of the dielectric media shown in Fig. 5.2.

5.4 THE "SPLIT" COAXIAL LINE

As in the case of the partially-filled coaxial line described in Section 5.3, two versions of the "split" coaxial line have been proposed, and both are illustrated in cross-section in Fig. 5.3. Neither version is seriously intended for use as a *transmission* line, i.e. primarily for the transport of energy, but both versions have been used in the design of baluns (i.e., transformers from BALanced to UNbalanced lines, as from coax. to microstrip). In addition, the version shown in Fig. 5.3(a) is in fairly wide use in the construction of coaxial slotted-line devices for the determination of electric-field distributions and magnitudes. An exact analysis of this structure, leading to a closed-form expression, has not, to the author's knowledge, been given. However, Duncan and Minerva [9] have given an extensive variational analysis of the problem which yields some rather complex, but nevertheless closed-form, formulae for upper and lower bounds on Z_0.

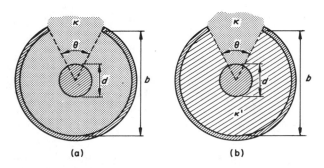

(a)   (b)

Fig. 5.3. Cross-sectional views of two versions of the "split" coaxial line

These are as follows:

Upper bound:

$$Z_0\sqrt{\kappa} = 59\cdot952\left\{\ln (b/d)+\frac{8}{(2\pi-\theta)^2}\sum_{n=1}^{\infty}\frac{\sin^2(n\theta/2)(1+\alpha N^2)^2}{n^3 X_n}\right\}$$

$$\text{ohm} \quad (5.4.1)$$

where

$$-\alpha = \frac{\sum_{n=1}^{\infty}\{N^2 \sin^2(n\theta/2)\}/n^3 X_n}{\sum_{n=1}^{\infty}\{N^4 \sin^2(n\theta/2)\}/n^3 X_n} \quad (5.4.2)$$

$$N^2 = n^2/(n^2-k^2) \quad (5.4.3)$$

$$k = \frac{1}{1-(\theta/2\pi)} \quad (5.4.4)$$

$$X_n = 1+\coth (n \ln b/d) \quad (5.4.5)$$

and d, b, θ are defined in Fig. 5.3(a).

Lower bound:

$$Z_0\sqrt{\kappa} = 59\cdot952\frac{\ln (b/d)}{1-\frac{4}{5}(\theta/2\pi)F(\theta)} \text{ ohm} \quad (5.4.6)$$

where

$$\frac{1}{F(\theta)} = \frac{4}{5}\left(\frac{\theta}{2\pi}\right)$$

$$(5.4.7)$$

$$+\frac{40}{\pi}\ln (b/d)\sum_{n=1}^{\infty}\frac{X_n}{(\theta/2)}\left\{\frac{A_n \cos (n\theta/2)-B_n \sin (n\theta/2)}{(n\theta/2)^4}\right\}^2$$

and

$$A_n = (n\theta/2)^3-6(n\theta/2) \quad (5.4.8)$$

$$B_n = 3(n\theta/2)^2-6. \quad (5.4.9)$$

Using measurements carried out on an "analogue" of the split coaxial line, consisting of the line cross-section drawn on resistive card, Duncan and Minerva claim to have shown that the arithmetic mean of the upper and lower bounding values of impedance agrees quite well with the measured results. However, since only three measured cases are quoted (for nominal coaxial line impedances of 50, 60, and 70 ohm), one of which is admitted to be inaccurate, this claim seems to be of somewhat dubious validity. From the results given in [9] it would appear that for practical application (in connection with slotted lines, for instance) a more accurate statement concerning accuracy of results would be as follows:

(a) For lines in which b/d corresponds to a 50 ohm standard (i.e. unslotted) line, the values obtained from equation (5.4.6) are in almost exact agreement with "measured" values.

(b) For lines in which b/d corresponds to a 75 ohm standard coaxial line, the values obtained by taking the arithmetic mean of equations (5.4.1) and (5.4.6) are in almost exact agreement with "measured" values.

In view of their possible interest to the measurements engineer, results calculated as described in paragraphs (a) and (b) above are tabulated in Table 5.1.

The "duo-dielectric" version of split coaxial line, shown in Fig. 5.3(b), was originally proposed by Hatsuda and Matsumoto [10], as a possible wide-band transformer from "standard" coaxial line to parallel-plate or microstrip transmission lines. The dielectric interfaces lead to complicated boundary conditions, and no analytic derivations of the line characteristics are known. Hatsuda and Matsumoto use relaxation techniques, in conjunction with a conformal transformation from cylindrical to rectangular coordinates, to evaluate numerically the characteristic impedance and propagation velocity. These are graphed for the particular cases $\kappa'/\kappa = 1$, $2 \cdot 26$, and $5 \cdot 0$, in [10] and [11]: the latter reference gives a more detailed description of the calculation procedure. Results have been checked experimentally for the cases $\kappa'/\kappa = 2 \cdot 26$ and $\theta = \pi/4$, $\pi/2$, $3\pi/4$, and show virtually exact agreement.

5.5 THE SLOTTED COAXIAL LINE

From the cross-sectional view shown in Fig. 5.4 it is apparent that this structure can be regarded as a further development or modification of the lines considered in the preceding two sections of this chapter. Its analysis is somewhat complicated by the fact that the greater symmetry of this structure allows two different TEM modes to propagate. The electric-field configurations of these are schematically displayed in Figs. 5.5(a) and 5.5(b). The former shows the "distorted" coaxial line mode which is of greatest interest in the present context, and the latter shows a "dipole radiator" type of mode which is of some interest in antenna technology. Both modes are discussed by Collin in [12], but detailed analysis is given only for a zero-thickness outer conductor.

TABLE 5.1 THE CHARACTERISTIC IMPEDANCE OF THE "SPLIT" COAXIAL LINE, AS CALCULATED FROM DUNCAN AND MINERVA'S FORMULAE [9], FOR THE TWO CASES IN WHICH THE UNSLOTTED COAXIAL LINE IMPEDANCE WOULD BE 50 ohm AND 75 ohm

Unslotted line characteristic impedance = $\theta°$	50 ohm $Z_0\sqrt{\kappa}$	75 ohm $Z_0\sqrt{\kappa}$
0	50·00	75·00
10	50·06	75·09
20	50·22	75·32
30	50·49	75·67
40	50·87	76·15
50	51·35	76·76
60	51·92	77·45
70	52·62	78·35
80	53·37	79·27
90	54·24	80·21
100	55·21	81·53
110	56·29	82·83
120	57·44	83·81
130	58·73	85·71
140	60·19	87·35
150	61·62	89·01
160	63·25	90·85
170	65·04	92·95
180	66·98	94·72
190	69·08	97·58
200	71·38	100·13
210	73·90	103·10
220	76·67	106·38
230	79·68	109·98
240	83·03	114·11
250	86·77	118·42
260	90·95	123·39
270	95·66	128·87
280	101·51	135·47
290	107·20	142·78
300	114·16	150·95
310	122·71	161·21
320	132·42	172·96
330	144·82	188·16
340	160·50	208·09

For this mode, Collin derives rather complex expressions for upper and lower bounds on the characteristic impedance, and assumes that the

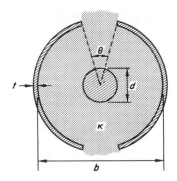

Fig. 5.4. Cross-section of the slotted coaxial line

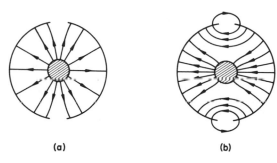

(a)   (b)

Fig. 5.5. Schematic illustration of the electric field distributions appropriate to the two possible TEM modes of propagation in the slotted coaxial line

arithmetic mean of these provides a close approximation to the "true" impedance. This mode is not of great interest in the present context, but for purposes of reference, Collin's formulae are given below:

Upper bound:

$$Z_0\sqrt{\kappa} = 59.952 \times \frac{32}{(\pi-\theta)^2} \sum_{n=1,3,5\cdots}^{\infty} \frac{\cos^2(n\theta/2)(1+\alpha N^2)^2}{n^3 X_n} \text{ ohm}$$

(5.5.1)

where

$$\alpha = \frac{\displaystyle\sum_{n=1,3,5\cdots}^{\infty} \{N^2 \cos^2(n\theta/2)\}/n^3 X_n}{\displaystyle\sum_{n=1,3,5\cdots}^{\infty} \{N^4 \cos^2(n\theta/2)\}/n^3 X_n}$$

(5.5.2)

$$N^2 = n^2/(n^2 - k^2)$$

(5.5.3)

$$k = \frac{2}{1-(\theta/\pi)}$$

(5.5.4)

$$X_n = 1 + \coth(n \ln b/d).$$

(5.5.5)

Note the very close resemblance between these equations and equations (5.4.1)–(5.4.5), which relate to the "split" coaxial line.

Lower bound:

$$Z_0\sqrt{\kappa} = \frac{296}{1.5 - \ln(\theta/2) + 4S(\theta)} \text{ ohm}$$

(5.5.6)

where

$$S(\theta) = \sum_{n=1,3,5\cdots}^{\infty} \frac{(X_n - 2)\sin^2(n\theta/2)}{n^3\theta^2}.$$

(5.5.7)

A more extensive treatment of the slotted coaxial line, for the case $t = 0$, is given by Smolarska [13], which includes analytic formulae for both possible modes of propagation. For the "distorted coaxial" mode of interest here, the formulae are very complex: and even though some simplification can be made by use of certain assumptions (not specified with any precision) which are stated to be valid in the majority of cases of likely interest, the resulting formulae are still somewhat complicated. The characteristic impedance is given by:

$$Z_0\sqrt{\kappa} \doteqdot 94.172 \frac{K(k)}{K'(k)} \text{ ohm}$$

(5.5.8)

where

$$k = \sqrt{\alpha}$$

(5.5.9)

and K, K', are the complete elliptic integrals defined and tabulated in Section 3.2. These formulae are simple enough, but Z_0 is related to the line dimensions and slot angle θ only through α, and the procedure for numerical evaluation is rather tedious, as follows.

Choose a value for α ($\leqslant 1$) and for b/d, then substitute into the following expression, and solve for γ:

$$\gamma\left\{\ln(b/d) - \frac{\pi}{2}\frac{K(k)}{K'(k)}\right\} + \sqrt{\gamma}E(k) + \frac{\pi}{2}(1+\alpha)\frac{K(k)}{K'(k)} - \ln(b/d) = 0.$$

(5.5.10)

Next, substitute the values of α and γ into the following expression, and solve for β:

$$\left(\frac{1+\beta}{\gamma+\beta}\right)K'(k) = \frac{\pi}{2\sqrt{\gamma}}.$$

(5.5.11)

Finally, θ can be calculated from

$$\pi - \theta = \frac{2\alpha\sqrt{\gamma}}{\gamma+\beta}\left\{\frac{(\beta+\alpha)}{\alpha}\mathbf{F}(\phi, k') - \frac{(\gamma+\beta)}{\gamma}\mathbf{\Pi}(\phi, -1, k')\right\}$$

(5.5.12)

where **F** and **Π** are, respectively, incomplete elliptic integrals of the 1st and 3rd kinds, and

$$\phi = \sin^{-1}\sqrt{\frac{\beta}{\alpha+\beta}}. \qquad (5.5.13)$$

A further brief, though detailed, discussion of the slotted coaxial line has been presented by Kogo [14]. Some graphs of Z_0 versus θ are given, for finite-thickness outer conductors ($t \neq 0$): specifically for $t/b = 0.1$, 0.5, and 1.0, relating to cases in which the "unslotted" coaxial line impedance would be 15, 25, and 50 ohm. Numerous experimental results are also displayed, which amply confirm the accuracy of the theoretical curves, and Kogo states that for $t = 0$, his results agree with Smolarska's.

5.6 THE POLYGONAL COAXIAL LINE

5.6.1 The Polygonal Coaxial Line with Cylindrical Centre Conductor

Regular polygonal coaxial lines with cylindrical inner conductors form a rather general class of possible transmission lines, of which two special cases have already been encountered in this book: the square slab-line of Section 4.2 (a regular N-gon with $N = 4$) and the ordinary coaxial line of Section 2.2 (which could be regarded as a regular N-gon with $N = \infty$).

These cases apart, it is difficult to imagine that polygonal lines can be of much practical value, except possibly in impedance transforming applications, or in applications requiring a high "packing density" of multiple transmission lines: in the latter case, the triangular ($N = 3$) or hexagonal ($N = 6$) lines could prove useful. The hexagonal version is shown in cross-section on Fig. 5.6.

However, the general case has been analysed by Seshagiri [15, 16], using a new approach called the "least-weighted-square invariance deformation" principle. As might be expected, the result-

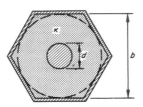

Fig. 5.6. The polygonal coaxial line with cylindrical centre conductor: cross-section of the special case $N = 6$ (hexagonal outer conductor)

ing formulae are quite complex and involved, and require computer facilities for detailed solution. The method is quite accurate as is shown by comparing results for $N = 4$ (i.e. square slab-line) with those already known from other analyses. (See Section 4.2.)

Table 5.2 lists values of $Z_0\sqrt{\kappa}$ as a function of the line dimensions, calculated from Seshagiri's formulae, for the cases $N = 3$–$N = 6$. The parameter d/b is the ratio between the diameter of the centre conductor and the diameter of the *inscribed* circle of the polygon.

5.6.2 The Polygonal Coaxial Line with Polygonal Centre Conductor

As with the preceding type of polygonal line, few practical applications spring to mind, apart from the one special case of $N = 4$, the square coaxial line already encountered in Section 3.3. Possibly, it could also be argued that the ordinary coaxial line (Section 2.2) is as much, or more, a special case of this form of polygonal line than of that described above in Section 5.6.1, bearing in mind the limiting process involved. Again, the most likely area of practical usage of these types of line is in the design of special impedance transformers, and with this in mind Green [17] has given numerical and semi-analytic derivations of the characteristic impedance.

Under the assumption that the corners of the polygon (or rather their capacitances) can be treated in isolation from the remainder of the structure, the total capacitance per unit length can be evaluated. Green quotes a formula derived by use of the Schwarz–Christoffel transformation [18], and this can be used to obtain the following expression for the characteristic impedance:

$$Z_0\sqrt{\kappa} = \frac{188.344\,(b/d-1)}{N\{\tan\,(\pi/N) + (b/d-1)X\}} \text{ ohm} \qquad (5.6.1)$$

where

$$X = \frac{1}{\pi}\left\{\psi\left(\frac{N+2}{2N}\right) - \psi(\tfrac{1}{2})\right\} \qquad (5.6.2)$$

$$\psi(x) = -\gamma + \sum_{N=1}^{\infty}\left\{\frac{x-1}{N(N+x-1)}\right\} \qquad (5.6.3)$$

$\gamma = 0.577216$ (Euler's constant), N is the number of sides to polygon, b is the diameter of *inscribed* circle of outer polygon, and d is the diameter of

TABLE 5.2 THE CHARACTERISTIC IMPEDANCE OF THE POLYGONAL COAXIAL LINE WITH CYLINDRICAL CENTRE CONDUCTOR, AS CALCULATED FROM SESHAGIRI'S FORMULAE (SEE TEXT, AND [15, 16]), FOR $N=$' 3, 4, 5, AND 6

d/b	$N=3$ $Z_0\sqrt{\kappa}$ (ohm)	$N=4$ $Z_0\sqrt{\kappa}$ (ohm)	$N=5$ $Z_0\sqrt{\kappa}$ (ohm)	$N=6$ $Z_0\sqrt{\kappa}$ (ohm)	d/b	$N=3$ $Z_0\sqrt{\kappa}$ (ohm)	$N=4$ $Z_0\sqrt{\kappa}$ (ohm)	$N=5$ $Z_0\sqrt{\kappa}$ (ohm)	$N=6$ $Z_0\sqrt{\kappa}$ (ohm)
0·05	184·32	183·28	182·58	181·64	0·50	46·24	44·95	44·30	43·43
0·06	173·35	172·31	171·62	170·69	0·51	45·08	43·77	43·12	42·25
0·07	164·07	163·04	162·36	161·43	0·52	43·93	42·61	41·96	41·08
0·08	156·02	155·01	154·34	153·42	0·53	42·82	41·48	40·82	39·95
0·09	148·93	147·92	147·26	146·34	0·54	41·72	40·37	39·71	38·83
0·10	142·58	141·58	140·92	140·01	0·55	40·64	39·28	38·61	37·73
0·11	136·84	135·85	135·20	134·29	0·56	39·59	38·21	37·54	36·65
0·12	131·60	130·61	129·97	129·06	0·57	38·55	37·16	36·48	35·60
0·13	126·78	125·79	125·15	124·26	0·58	37·53	36·12	35·44	34·56
0·14	122·32	121·33	120·70	119·80	0·59	36·53	35·11	34·42	33·54
0·15	118·16	117·18	116·55	115·66	0·60	35·55	34·11	33·42	32·53
0·16	114·28	113·30	112·67	111·78	0·61	34·58	33·13	32·44	31·55
0·17	110·63	109·65	109·03	108·14	0·62	33·63	32·16	31·47	30·58
0·18	107·19	106·21	105·59	104·71	0·63	32·69	31·22	30·52	29·62
0·19	103·94	102·96	102·34	101·46	0·64	31·77	30·28	29·58	28·68
0·20	100·86	99·87	99·26	98·38	0·65	30·86	29·36	28·65	27·76
0·21	97·93	96·94	96·33	95·45	0·66	29·97	28·46	27·74	26·85
0·22	95·13	94·14	93·53	92·66	0·67	29·09	27·56	26·85	25·95
0·23	92·47	91·47	90·86	89·99	0·68	28·22	26·69	25·97	25·07
0·24	89·91	88·91	88·30	87·43	0·69	27·36	25·82	25·10	24·20
0·25	87·46	86·45	85·85	84·98	0·70	26·52	24·97	24·24	23·34
0·26	85·11	84·10	83·50	82·63	0·71	25·69	24·12	23·39	22·49
0·27	82·85	81·83	81·23	80·36	0·72	24·86	23·29	22·56	21·66
0·28	80·68	79·65	79·05	78·18	0·73	24·05	22·47	21·74	20·84
0·29	78·58	77·54	76·94	76·08	0·74	23·25	21·66	20·93	20·02
0·30	76·55	75·50	74·91	74·04	0·75	22·45	20·86	20·13	19·22
0·31	74·59	73·54	72·94	72·07	0·76	21·66	20·07	19·33	18·43
0·32	72·70	71·63	71·03	70·17	0·77	20·89	19·29	18·55	17·65
0·33	70·86	69·79	69·19	68·32	0·78	20·12	18·52	17·78	16·88
0·34	69·08	68·00	67·40	66·53	0·79	19·35	17·75	17·02	16·12
0·35	67·36	66·26	65·66	64·79	0·80	18·59	16·99	16·26	15·37
0·36	65·68	64·57	63·97	63·11	0·81	17·84	16·25	15·52	14·63
0·37	64·05	62·93	62·33	61·46	0·82	17·09	15·50	14·78	13·89
0·38	62·47	61·34	60·73	59·86	0·83	16·35	14·77	14·04	13·16
0·39	60·92	59·78	59·17	58·31	0·84	15·61	14·03	13·32	12·44
0·40	59·42	58·27	57·65	56·79	0·85	14·87	13·31	12·60	11·73
0·41	57·96	56·79	56·17	55·31	0·86	14·14	12·59	11·88	11·02
0·42	56·53	55·35	54·73	53·87	0·87	13·40	11·87	11·17	10·32
0·43	55·14	53·95	53·32	52·46	0·88	12·66	11·15	10·47	9·63
0·44	53·78	52·57	51·95	51·08	0·89	11·92	10·43	9·76	8·94
0·45	52·45	51·23	50·60	49·74	0·90	11·17	9·72	9·06	8·25
0·46	51·15	49·92	49·29	48·42	0·91	10·42	9·00	8·36	7·57
0·47	49·88	48·64	48·00	47·13	0·92	9·65	8·27	7·66	6·89
0·48	48·64	47·38	46·74	45·87	0·93	8·87	7·54	6·95	6·20
0·49	47·43	46·15	45·51	44·64	0·94	8·07	6·80	6·23	5·52

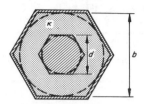

Fig. 5.7. The polygonal coaxial line with polygonal centre conductor: cross-section of the special case $N = 6$ (the hexagonal coaxial line)

inscribed circle of inner polygon. (See Fig. 5.7, which illustrates the special case $N = 6$.)

These formulae have been used to evaluate the data listed in Table 5.3, which gives $Z_0\sqrt{\kappa}$ as a function of d/b for N-gonal lines, where $N = 3, 4, 5,$ and 6. The data for $N = 4$ may be compared with Cockcroft's data for the square coaxial line which are tabulated in Table 3.2: it can be shown that equation (5.6.1) is identical with Cockcroft's formula, for the case $N = 4$. Results obtained are accurate for $d/b > 0.25$.

5.7 THE MULTI-FIN COAXIAL LINE

This unusual structure consists of a cylindrical outer conductor enclosing a centre conductor which is composed of N radial fins. The case $N = 6$ is illustrated in Fig. 5.8, assuming fins of negligible thickness.

An analysis, based upon the use of conformal transformations, has been given by Boyd [19]. It is shown that the characteristic impedance of any multi-fin line can be expressed in terms of that for the 2-fin case, which is in fact identical with the coaxial stripline described and analyzed in Section 4.6.

Boyd's formulae can be written as follows:

$$Z_{0N} = \frac{2}{N}Z_{02} \text{ ohm} \qquad (5.7.1)$$

when

$$\frac{2r_N}{b_N} = \left(\frac{W}{b}\right)^{2/N} \qquad (5.7.2)$$

where r_N is the radius of fin in N-fin line, b_N is the diameter of outer conductor, Z_{0N} is the characteristic impedance of line with N symmetric radial fins, Z_{02} is the characteristic impedance of 2-fin line, i.e. the coaxial stripline of Section 4.6, and $W = 2r_2$ is the width of strip in coaxial stripline.

From equations (4.6.1) and (4.6.2) we obtain

$$Z_{02}\sqrt{\kappa} = 94 \cdot 172 \frac{K'(x)}{K(x)} \text{ ohm} \qquad (5.7.3)$$

where

$$x = \frac{2W/b}{1 + (W/b)^2} \qquad (5.7.4)$$

See Fig. 4.8 for graph of Z_{02} versus W/b.

As Boyd points out [19], it might be expected that for large N, Z_{0N} will approach the value of Z_0 appropriate to the "ordinary" coaxial line (Chapter 2), and this appears to be confirmed by the graphs given in the reference cited.

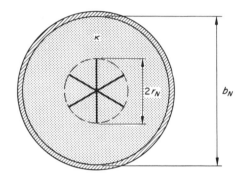

Fig. 5.8. Cross-section of the multi-fin coaxial line, for the special case of 6 fins

TABLE 5.3 THE CHARACTERISTIC IMPEDANCE OF THE POLYGONAL COAXIAL LINE WITH POLYGONAL CENTRE CONDUCTOR, AS CALCULATED FROM GREEN'S FORMULA (SEE TEXT, AND [18])

d/b	$N=3$ $Z_0\sqrt{\kappa}$ (ohm)	$N=4$ $Z_0\sqrt{\kappa}$ (ohm)	$N=5$ $Z_0\sqrt{\kappa}$ (ohm)	$N=6$ $Z_0\sqrt{\kappa}$ (ohm)	d/b	$N=3$ $Z_0\sqrt{\kappa}$ (ohm)	$N=4$ $Z_0\sqrt{\kappa}$ (ohm)	$N=5$ $Z_0\sqrt{\kappa}$ (ohm)	$N=6$ $Z_0\sqrt{\kappa}$ (ohm)
0·05	145·10	141·80	137·00	133·10	0·50	30·28	36·80	39·11	40·10
0·06	138·87	137·18	133·06	129·55	0·51	29·28	35·66	37·94	38·93
0·07	133·04	132·75	129·27	126·11	0·52	28·31	34·55	36·79	37·78
0·08	127·57	128·52	125·61	122·78	0·53	27·36	33·46	35·67	36·65
0·09	122·42	124·46	122·07	119·55	0·54	26·44	32·40	34·57	35·54
0·10	117·57	120·57	118·66	116·43	0·55	25·54	31·36	33·49	34·45
0·11	113·00	116·83	115·37	113·40	0·56	24·66	30·34	32·43	33·38
0·12	108·67	113·25	112·18	110·46	0·57	23·80	29·34	31·40	32·34
0·13	104·58	109·80	109·09	107·60	0·58	22·97	28·36	30·38	31·30
0·14	100·69	106·48	106·11	104·83	0·59	22·15	27·40	29·38	30·29
0·15	97·01	103·28	103·22	102·13	0·60	21·36	26·46	28·40	29·30
0·16	93·50	100·20	100·42	99·52	0·61	20·58	25·54	27·43	28·32
0·17	90·17	97·24	97·71	96·97	0·62	19·82	24·64	26·49	27·36
0·18	86·99	94·37	95·07	94·49	0·63	19·08	23·76	25·56	26·41
0·19	83·95	91·61	92·52	92·08	0·64	18·35	22·89	24·65	25·48
0·20	81·06	88·94	90·04	89·74	0·65	17·64	22·04	23·75	24·57
0·21	78·29	86·36	87·63	87·45	0·66	16·95	21·20	22·87	23·67
0·22	75·64	83·86	85·29	85·23	0·67	16·27	20·39	22·01	22·79
0·23	73·10	81·45	83·02	83·06	0·68	15·61	19·58	21·15	21·92
0·24	70·66	79·11	80·81	80·95	0·69	14·96	18·80	20·32	21·06
0·25	68·32	76·84	78·66	78·89	0·70	14·32	18·02	19·50	20·22
0·26	66·08	74·65	76·56	76·88	0·71	13·70	17·26	18·69	19·39
0·27	63·92	72·52	74·53	74·92	0·72	13·09	16·52	17·90	18·57
0·28	61·85	70·46	72·54	73·00	0·73	12·50	15·78	17·11	17·77
0·29	59·85	68·45	70·61	71·14	0·74	11·91	15·06	16·35	16·98
0·30	57·93	66·51	68·73	69·31	0·75	11·34	14·36	15·59	16·20
0·31	56·07	64·62	66·89	67·53	0·76	10·78	13·66	14·85	15·43
0·32	54·28	62·78	65·10	65·79	0·77	10·22	12·98	14·11	14·68
0·33	52·55	61·00	63·35	64·08	0·78	9·68	12·31	13·39	13·94
0·34	50·88	59·26	61·65	62·42	0·79	9·16	11·65	12·68	13·20
0·35	49·27	57·57	59·99	60·79	0·80	8·64	11·00	11·99	12·48
0·36	47·71	55·93	58·36	59·20	0·81	8·13	10·37	11·30	11·77
0·37	46·20	54·33	56·78	57·65	0·82	7·63	9·74	10·62	11·07
0·38	44·74	52·77	55·23	56·12	0·83	7·14	9·12	9·95	10·38
0·39	43·33	51·25	53·72	54·63	0·84	6·65	8·52	9·30	9·70
0·40	41·96	49·77	52·24	53·17	0·85	6·18	7·92	8·65	9·03
0·41	40·63	48·33	50·79	51·74	0·86	5·72	7·33	8·02	8·37
0·42	39·34	46·92	49·38	50·34	0·87	5·26	6·75	7·39	7·71
0·43	38·09	45·55	48·00	48·97	0·88	4·81	6·19	6·77	7·07
0·44	36·88	44·21	46·64	47·63	0·89	4·37	5·63	6·16	6·44
0·45	35·70	42·90	45·32	46·31	0·90	3·94	5·07	5·56	5·81
0·46	34·55	41·62	44·02	45·02	0·91	3·52	4·53	4·97	5·19
0·47	33·44	40·38	42·76	43·75	0·92	3·10	4·00	4·38	4·59
0·48	32·36	39·16	41·51	42·51	0·93	2·69	3·47	3·81	3·99
0·49	31·30	37·97	40·30	41·29	0·94	2·28	2·95	3·24	3·39

REFERENCES

1. Smythe, W. R. *Static and Dynamic Electricity*.
   p. 466, McGraw-Hill, New York (1950).
2. *Ibid.*, p. 102.
3. King, M. J. and Wiltse, J. C. "Coaxial Transmission Lines of Elliptical Cross-Section", *Trans. I.R.E.*
   AP-9, pp. 116–118 (Jan., 1961).
4. Mahapatra, S. "Coaxial Lines", *Electronic and Radio Engr.*
   35, 63–67 (Feb., 1958).
5. Angelakos, D. J. "A Coaxial Line filled with Two Non-Concentric Dielectrics", *Trans. I.R.E.*
   MTT-2, pp. 39–44 (July, 1954).
6. Sullivan, D. J. and Parkes, D. A. "Stepped Transformers for Partially-Filled Transmission Lines", *Trans. I.R.E.*
   MTT-8, pp. 212–217 (March, 1960).
7. Smorgonski, S. Y. "Determination of the types of waves in a Coaxial Cable with Elliptical Conductors", *Radio Engng and electron. Phys.*
   pp. 85–90 (Jan., 1964).
8. Bräckelmann, W. "Die Grenzfrequenzen von höheren Wellentypen in Koaxialkabel mit elliptischem Querschnitt", *A.E.Ü.*
   21, 421–426 (Aug., 1967).
9. Duncan, J. W. and Minerva, V. P. "100:1 Bandwidth Balun Transformer", *Proc. I.R.E.*
   48, 156–164 (Feb., 1960).
10. Hatsuda, T. and Matsumoto, T. "Computation of Impedance of Partially Filled and Slotted Coaxial Line", *Trans. I.E.E.E.*
    MTT-15, pp. 643–644 (Nov., 1967).
11. Hatsuda, T. *et al.* "Computation of Characteristic Impedance of Inhomogeneous Slotted Coaxial Lines by Relaxation Method", *Electronics Communs. Jap.*
    51-5, 4, pp. 91–99 (April, 1968).
12. Collin, R. E. "The Characteristic Impedance of a Slotted Coaxial Line", *Trans. I.R.E.*
    MTT-4, pp. 4–8 (Jan., 1956).
13. Smolarska, J. "Characteristic Impedances of the Slotted Coaxial Line", *Trans. I.R.E.*
    MTT-6, pp. 161–166 (April, 1958).
14. Kogo, H. "Characteristic Impedance of Split Coaxial Line", *Trans. I.R.E.*
    MTT-7, pp. 393–394 (July, 1959).
15. Seshagiri, N. "Least-Weighted-Square Method for Analysis and Synthesis of Transmission Lines", *Trans. I.E.E.E.*
    MTT-15, pp. 494–503 (Sept., 1967).
16. Seshagiri, N. "Regular Polygon Circular Coaxial Transmission Line", *Proc. I.E.E.E.*
    53, 1749–1750 (Nov., 1965).
17. Green, H. E. "The Numerical Solution of Transmission Line Problems", *Advances in Microwaves.*
    Vol. 2, pp. 352–354, Academic Press, New York (1967).
18. Cunningham, J. *Complex Variable Methods.*
    Van Nostrand, Lond. (1965).
19. Boyd, C. R. "Characteristic Impedance of Multi-fin Transmission Lines", *Trans. I.E.E.E.*
    MTT-15, pp. 487–488 (Aug., 1967).

6

COUPLED TRANSMISSION LINES

6.1 INTRODUCTION

When two (or more) unshielded transmission lines are located in close proximity to one another, they become electromagnetically coupled via their associated electric and magnetic fields; particularly if the line axes are parallel. In many situations, such coupling is highly undesirable since it gives rise to "cross-talk", background noise, distortion, loss of energy, and other detrimental effects. These can usually be avoided, or at least reduced to an acceptable level, by the provision of shielding conductors: either separately, or as an integral part of the transmission line structure, e.g. the outer conductor of the coaxial line (Chapter 2) which, if sufficiently thick, provides a highly efficient electromagnetic screen.

However, the present text is rather more concerned with those cases in which coupling between adjacent lines is desired, as in the practical realization of microwave devices such as frequency filters (for the selection and/or suppression of specified bands of frequencies), directional couplers (for power-dividing or power-sampling applications), phase changers, group delay equalizers, etc.

In designing such devices, the desired degree of coupling is usually known or specified, and it is required to determine the line dimensions and spacing necessary to achieve this degree of coupling. This can be effected via a knowledge of the even-mode and odd-mode characteristic impedances, Z_{0e} and Z_{0o}, of a pair of coupled lines, which were defined in Section 1.5.2.

It has been shown, for instance [1, 2], that when two (matched) transmission lines are electromagnetically coupled by following parallel paths a distance "S" apart, the power P_2 coupled into the second line when power P_1 is incident in the first line is given by

$$P_2 = \frac{P_1 k^2 \sin^2\theta}{1 - k^2 \cos^2\theta} \qquad (6.1.1)$$

where θ is the electrical length of the parallel coupled sections of line, and k is the "coupling factor" defined by

$$k = \frac{Z_{0e} - Z_{0o}}{Z_{0e} + Z_{0o}}. \qquad (6.1.2)$$

This chapter is therefore devoted to formulae, tables, and graphs, of Z_{0e} and Z_{0o} as functions of the line parameters, for the coupled microwave transmission lines in common and not-so-common, use.

It will be apparent, from a consideration of the physical configurations of the various line structures already treated in the previous chapters, that the number of different possible types of coupled lines will be considerably smaller than the number of different types of single lines. It is therefore considered appropriate that they should be treated in separate sections, but within a single chapter.

6.2 LINES WITH COUPLED CONDUCTORS OF CIRCULAR CROSS-SECTION

6.2.1 Coupled Slab-Lines

The single slab-line in its various forms has been discussed in Chapter 4, and the coupled configuration now under discussion (see Fig. 6.1 for cross-sectional diagram) has found wide application in the construction of filters [3, 4] and directional couplers [5]. In designing these devices, it is necessary to be able to determine Z_{0o} and Z_{0e} as functions of the line parameters, and one of Frankel's [6] characteristic impedance formulae can easily be written as an expression for Z_{0o}, as follows:

$$Z_{0o}\sqrt{\kappa} = 59 \cdot 952 \ln\left\{\frac{4b}{\pi d} \tanh\left(\frac{\pi S}{2b}\right)\right\} \text{ ohm.} \qquad (6.2.1)$$

However, no accompanying formula for Z_{0e} can be obtained from Frankel's work and since equation (6.2.1) is valid only for $d \ll b$ and $S \gg d$, the result is really only of historical interest.

Somewhat more accurate, yet simple, formulae have been derived by R. C. Honey, and quoted without proof in a paper by Jones and Bolljahn [1].

$$Z_{0e} - Z_{0o} \doteq \frac{119 \cdot 904}{\sqrt{\kappa}} \ln \coth\left(\frac{\pi S}{2b}\right) \text{ ohm} \qquad (6.2.2)$$

$$Z_{0e} + Z_{0o} \doteq \frac{119 \cdot 904}{\sqrt{\kappa}} \ln \coth\left(\frac{\pi d}{4b}\right) \text{ ohm.} \qquad (6.2.3)$$

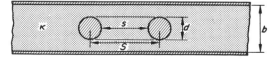

Fig. 6.1. Cross-section of coupled slab lines. Note that S is the centre-to-centre spacing of cylindrical conductors, and s is their "edge-to-edge" spacing

In a later work, Bolljahn and Matthaei [7] state that these formulae are believed to be accurate provided that $d/b < 0 \cdot 25$ and $S/b > 3d/b$; and they have been used with considerable success by Vadopalas [8] in the design of comb-line and interdigital filters.

Note in particular, that when d/b is small (e.g. $d/b \ll 0 \cdot 25$), the above equations can be solved simultaneously to yield:

$$Z_{0e}\sqrt{\kappa} \doteq 59 \cdot 952 \ln\left\{\frac{4b}{\pi d} \coth\left(\frac{\pi S}{2b}\right)\right\} \text{ ohm} \qquad (6.2.4)$$

$$Z_{0o}\sqrt{\kappa} \doteq 59 \cdot 952 \ln\left\{\frac{4b}{\pi d} \tanh\left(\frac{\pi S}{2b}\right)\right\} \text{ ohm.} \qquad (6.2.5)$$

These are precisely the formulae presented by Levy [10], and for which it is stated that they are believed to be accurate for $d/b \leqslant 0 \cdot 6$. This seems unlikely, since they were derived above under the assumption $d/b \ll 0 \cdot 25$: note also that equation (6.2.5) is identical with Frankel's formula (see equation (6.2.1)), which also is valid only under the same limited conditions.

Although the formulae quoted above are extremely useful for design purposes (within their stated limitations) because of their simple, closed-form nature, the "classic" source of data on coupled slab-lines is undoubtedly Cristal's paper [9]. This presents, in detailed graphical form, the results of an extensive and highly-accurate numerical analysis of a multiple coupled-rod structure. The data obtained consists of interelectrode capacitances C_m and C_e, as functions of the line parameters, where C_m is the capacitance between an adjacent pair of conductors, and C_e is the total capacitance between one conductor and the two ground planes.

It must be emphasized here that the capacitance values thus obtained relate only to even-mode and odd-mode excitation of a pair of coupled conductors *which constitute part of an infinite array of identical, equally spaced conductors*. As presented, the data given can be directly used for the design of, for example, interdigital and comb-line filters. However, for the design of parallel-coupled filters, or directional couplers, which are constituted of *single pairs* of identical conductors, a modified technique is required to determine the correct interconductor capacitances in terms of line dimensions. This is fairly easy to reason out, but full details are also given (with reference to the end conductor in a given array) in [4] and [9].

For convenience in performing such calculations, Nicholson [4] has redrawn Cristal's graphs, and presented them in the form of a single multi-scale chart, which is reproduced here (by kind permission of the I.E.R.E.) as Fig. 6.2.

When the capacitances appropriate to a particular structure and mode of excitation have been correctly evaluated, the even- and odd-mode impedances are given by the following formulae:

$$Z_{0e}\sqrt{\kappa} = \frac{376\cdot687}{C'_e} \text{ ohm} \tag{6.2.6}$$

$$Z_{0o}\sqrt{\kappa} = \frac{376\cdot687}{C'_e+4C'_m} \text{ ohm} \tag{6.2.7}$$

where

$$C'_e = C_e/c_0\kappa \quad \text{and} \quad C'_m = C_m/c_0\kappa.$$

Although extensively used for filter and coupler design, the graphical data presented suffers from the severe disadvantage that a fair amount of "trial and error" is required in order to achieve satisfactory results. In order to overcome this, and also to expedite the development of completely computerized design procedures, a number of workers have resorted to curve-fitting techniques and numerical interpolation, utilizing Cristal's data. The accuracy of such design procedures is obviously strongly dependent upon the curve-fitting and interpolation processes employed.

The design and synthesis problem has also been treated by Conning [11], who also uses Cristal's capacitance graphs as a starting point, but employs curve-fitting techniques to derive semi-empirical formulae for C'_e and C'_m. These obviously offer considerable advantages for use in computerized design procedures, provided that the attainable accuracy is adequate for the purpose in mind. Conning claims that the formulae fit Cristal's curves to within ±2 per cent for $s/b \geqslant 0\cdot2$ and $0\cdot1 < d/b < 0\cdot8$: it should, of course, also be borne in mind that the curves themselves, although highly accurate, are not exact, so that the final residual error may well be greater, or less, than the nominal 2 per cent. The empirical formulae are:

$$C'_m = 3\cdot75\left(\frac{d}{b}\right)^{0\cdot42}\left\{e^{-\pi s/b}+\left(1-0\cdot036\frac{b}{d}\right)E\right\} \tag{6.2.8}$$

where

$$E = \left(1-\frac{1}{2}\frac{d^2}{b^2}\right)e^{-3\pi s/b}+e^{-25s/b}$$

$$C'_e+2C'_m = \frac{2\pi}{\ln(4b/\pi d)+\frac{2}{9}(\pi d/4b)^4}+14\left(\frac{d}{b}\right)^{0\cdot7}e^{-8s/b}. \tag{6.2.9}$$

These expressions can be used with equations (6.2.6) and (6.2.7) to obtain closed-form formulae for Z_{0e} and Z_{0o}.

6.2.2 Split-coupled Slab-Lines

This rather unusual coupled-line structure is simply obtained by splitting the centre conductor of a single slab-line (Section 4.4) vertically along a diameter, and separating the two halves: thus forming two separate, but coupled, transmission lines. A cross-sectional view is shown in Fig. 6.3, which serves to define the dimensional parameters.

This form of coupling is useful for the design of directional couplers, particularly for tight coupling (3 dB or thereabouts) and large bandwidth. The main advantages, as stated by Pon [12] are the large coupling area, the small cross-section (allowing high-frequency operation), and reasonable conductor size (for ease of fabrication). The main disadvantages are, probably, the difficulty of maintaining the necessary parallel alignment, and the problems arising from the necessity of transforming from the semi-circular centre-conductor cross-section to that of "standard" slab, coaxial, or triplate line.

For very tight coupling ($s \ll d$), the required impedance formulae have been derived by Mc-Dermott, and quoted by Levy [10], as follows:

$$Z_{0e}\sqrt{\kappa} \doteq 119\cdot9\ln\left(\frac{4b}{\pi d}\right)\text{ ohm} \tag{6.2.10}$$

$$Z_{0o}\sqrt{\kappa} \doteq \frac{Z_{0e}\sqrt{\kappa}}{1+0\cdot0531Z_{0e}\sqrt{\kappa}(d/s+0\cdot441)}\text{ ohm.} \tag{6.2.11}$$

It should perhaps be mentioned that [10] contains a slightly ambiguous diagram, in which the dimension $d+s$ is misleadingly indicated as d.

The value of $Z_{0e}\sqrt{\kappa}$ obtained from equation (6.2.10) is in fact precisely double the value of $Z_0\sqrt{\kappa}$ for the unscreened slab-line obtained from Frankel's formula (equation (4.4.2)), which is valid for small d/b. Somewhat more general formulae, derived under the assumption that the fringing capacitances arising at the diametral edges are

Coupling capacitance $\dfrac{C_m}{\kappa\epsilon_0} = C_m'$

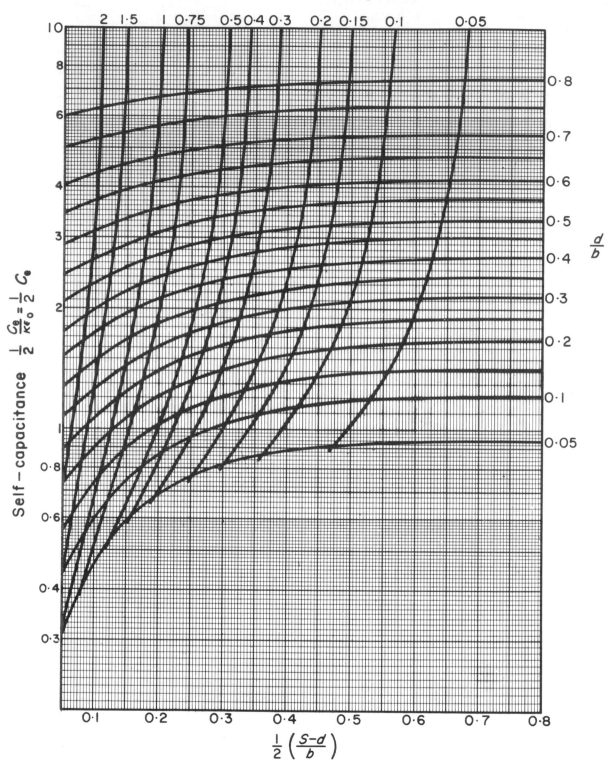

Fig. 6.2. Graphs of interelectrode capacitances for coupled slab-lines, as calculated by Cristal [9]. (Courtesy of B. Nicholson, and the I.E.R.E.)

approximately the same as those arising at the edges of a rectangular-section inner conductor, have been presented by Pon [12]:

$$Z_{0e}\sqrt{\kappa} = \frac{376 \cdot 687}{C_e + 2C_{fe}} \text{ ohm} \qquad (6.2.12)$$

$$Z_{0o}\sqrt{\kappa} = \frac{376 \cdot 687}{C_e + 2C_{fo}} \text{ ohm} \qquad (6.2.13)$$

where

$$C_e = \frac{\pi}{\ln(4b/\pi d)} \qquad (6.2.14)$$

$$C_{fe} = \frac{2}{\pi}\left(0 \cdot 89 + 2 \cdot 14\frac{d}{b}\right)\ln\left(1 + \tanh\frac{\pi s}{2b}\right) \qquad (6.2.15)$$

$$C_{fo} = \frac{2}{\pi}\ln\left(1 + \coth\frac{\pi s}{2b}\right) + \frac{d}{s}\left(1 - \frac{s}{2b}\right). \qquad (6.2.16)$$

The accuracy of these semi-empirical formulae has been adequately verified in the design of experimental 3dB couplers [12].

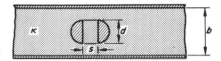

Fig. 6.3. Cross-section of the split-coupled slab-lines

6.3 LINES WITH COUPLED CONDUCTORS OF RECTANGULAR CROSS-SECTION

6.3.1 Edge-coupled Zero-thickness Triplate Stripline

As in the case of uncoupled triplate, the earliest, and probably best-known, analysis was given by

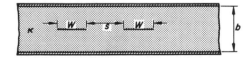

Fig. 6.4. Cross-section of edge-coupled zero-thickness triplate striplines

Cohn [13]: the line cross-section with zero-thickness strips (see Fig. 6.4) can be analyzed exactly by the use of conformal transformation, and Cohn quotes the following formulae:

$$Z_{0e}\sqrt{\kappa} = 29 \cdot 976\pi \frac{K'(k_e)}{K(k_e)} \text{ ohm} \qquad (6.3.1)$$

$$Z_{0o}\sqrt{\kappa} = 29 \cdot 976\pi \frac{K'(k_0)}{K(k_0)} \text{ ohm} \qquad (6.3.2)$$

where K, K' are complete elliptic integrals of the 1st kind (see Section 3.2), and

$$k_e = \tanh\left(\frac{\pi W}{2b}\right)\tanh\left(\frac{\pi}{2} \cdot \frac{W+s}{b}\right) \qquad (6.3.3)$$

$$k_0 = \tanh\left(\frac{\pi W}{2b}\right)\coth\left(\frac{\pi}{2} \cdot \frac{W+s}{b}\right) \qquad (6.3.4)$$

As pointed out in Chapter 3, the strip thicknesses used in microwave printed circuits are often sufficiently small so that circuit components can be designed quite accurately by the use of zero-thickness formulae. In designing filters and directional couplers, the desired performance will define the values of Z_{0e} and Z_{0o} required, and the problem then is to determine the values of W/b and s/b which will realize the specified impedances. Equations (6.3.3) and (6.3.4) can be inverted [13] to yield the following expressions for the line parameters in terms of k_e and k_0:

$$\frac{W}{b} = \frac{2}{\pi}\tanh^{-1}\sqrt{k_e k_0} \qquad (6.3.5)$$

$$\frac{s}{b} = \frac{2}{\pi}\tanh^{-1}\left(\frac{\sqrt{k_e}}{1-k_e} \cdot \frac{1-k_0}{\sqrt{k_0}}\right). \qquad (6.3.6)$$

Given Z_{0e} and Z_{0o}, k_e and k_0 can quickly be determined from equations (6.3.1) and (6.3.2) by use of the elliptic integral tables given in Section 3.2; the required coupled line dimensions then immediately follow from equations (6.3.5) and (6.3.6).

This design synthesis problem has also been solved by Singletary [14], using the results of Cohn's analysis to derive direct relationships between line dimensions and required impedances:

$$\frac{s}{b} = -\frac{1}{\pi}\ln\left\{\tanh 47 \cdot 086\pi\left(\frac{1}{Z_{0o}\sqrt{\kappa}} - \frac{1}{Z_{0e}\sqrt{\kappa}}\right)\right\} \qquad (6.3.7)$$

$$\frac{W}{b} = \frac{94 \cdot 172}{Z_{0e}\sqrt{\kappa}} - \frac{s}{2b} + \frac{1}{\pi}\ln\left\{\frac{1}{2}\cosh\frac{\pi s}{2b}\right\} \qquad (6.3.8)$$

The problem of two coupled strips of zero-thickness and unequal width has been considered by Sato and Ikeda [15] and some extremely complex expressions are derived for the charge and potential of each strip, from which the capacitances, and hence the impedances, could (in principle!) be obtained. However, no explicit formulae for these parameters are given.

6.3.2 Edge-coupled Finite-thickness Triplate Stripline

Although many circuit synthesis problems can be solved by design procedures based upon the use of zero-thickness-strip impedance data, there are many other applications in which the use of strip conductors of appreciable thickness is desirable, and this necessitates the use of more accurate design data, because the non-zero strip thickness considerably modifies the various capacitances involved in even- and odd-mode impedance calculations.

This finite-thickness, or "thick-strip" problem was first treated by Cohn [13], as an extension of the zero-thickness case. He derived correction terms to be applied to the exact zero-thickness-strip formulae, and a similar approach was also adopted by Horgan [16].

However, a more direct approach to the analysis of the coupled thick-strip structure as illustrated in Fig. 6.5 was made by Getsinger [17], based as

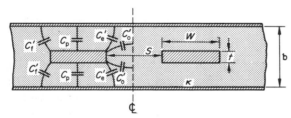

Fig. 6.5. Cross-section of edge-coupled finite-thickness triplate striplines, showing interelectrode capacitances as defined by Getsinger [17]. See text for details

usual upon the application of conformal mapping techniques involving the use of the Schwarz–Christoffel transformation. The formulae obtained are extremely complex functions of elliptic integrals and Jacobian functions (see [17]), and Getsinger has therefore presented his results in graphical form. The graphs are of doubtful accuracy for small values of s/b (tightly coupled strips) and Gupta [41] has recalculated and extended them for $0.02 \leqslant s/b \leqslant 0.2$. Getsinger's and Gupta's graphs are reproduced (by kind permission of the IEEE), in slightly modified form,

on Figs. 6.6 and 6.7. They have been extensively used in the practical design of filters, directional couplers, delay equalizers and similar devices.

Having obtained the relevant (normalized) fringing capacitances from these graphs, the even- and odd-mode characteristic impedances of finite-thickness, edge-coupled triplate stripline are given by:

$$Z_{0e}\sqrt{\kappa} = \frac{188{\cdot}344}{C'_p + C'_f + C'_e} \text{ ohm} \qquad (6.3.9)$$

$$Z_{0o}\sqrt{\kappa} = \frac{188{\cdot}344}{C'_p + C'_f + C'_o} \text{ ohm} \qquad (6.3.10)$$

In these formulae, C'_p is the simple (normalized) parallel-plate capacitance given by:

$$C'_p = \frac{2W/b}{1 - t/b} \qquad (6.3.11)$$

The fringing capacitance C'_f is independent of s/b, and is given by the asymptotic values of C_e or C_o ($s/b \to \infty$). To sufficient accuracy for most purposes, C_f is given by $\frac{1}{2}(C'_e + C'_o)$ for $s/b = 1{\cdot}5$, using the curves relating to the appropriate value of t/b on Figs. 6.7(a) and (b).

For design applications, in which it is usually required to determine the values of W/b and s/b which will realize specified values of $Z_{0e}\sqrt{\kappa}$ and $Z_{0o}\sqrt{\kappa}$ for a given t/b, the formulae can be cast into a more readily usable form. From Getsinger's paper [17] we obtain (*using different notation*):

$$\Delta C = 188{\cdot}344 \left(\frac{1}{Z_{0o}\sqrt{\kappa}} - \frac{1}{Z_{0e}\sqrt{\kappa}} \right) \qquad (6.3.12)$$

and

$$\frac{W}{b} = \frac{1}{2}\left(1 - \frac{t}{b}\right)\left(\tfrac{1}{2}C'_{0e} - C'_e - C'_f\right) \qquad (6.3.13)$$

where

$$\tfrac{1}{2}C'_{0e} = C'_p + C'_e + C'_f. \qquad (6.3.14)$$

The procedure is as follows: using the specified values of Z_{0e}, Z_{0o} and κ, ΔC is evaluated from equation (6.3.12). Since t/b will also be specified, the appropriate value of s/b can be determined from the graphs of Fig. 6.6(b). Hence, corresponding values of C'_e and C'_f can be read off from Figs. 6.6 and 6.7. Next, $\frac{1}{2}C'_{0e}$ can be calculated, from equations (6.3.14) and (6.3.9), and then all parameters for insertion into equation (6.3.13) are known, thus determining W/b.

The case where side walls are present, so that the coupled strips are completely shielded from the external surroundings, is treated by Sato and Ikeda [15], but no directly usable results are given: all formulae are expressed in terms of a complex integral, which is left unevaluated, and is not, to the author's knowledge, expressible in terms of known or tabulated functions.

A procedure whereby the even- and odd-mode impedances of a pair of thick coupled strips of *unequal* width can be evaluated, using Getsinger's data for equal-width strips (Figs. 6.6 and 6.7), is outlined in [18].

The cut-off frequencies of higher-order modes in edge-coupled triplate structures have been calculated by Ilenburg and Pregla [19]: it is shown that in the structure of Fig. 6.5, only TE modes can propagate, and that their cut-off frequencies are higher than in the case of a completely shielded structure (i.e. with side walls).

A method of calculating the losses in coupled triplate structures, using Getsinger's or Gupta's charts, has been described by Horton [42].

6.3.3 Broadside-coupled (Horizontal) Stripline

This configuration, illustrated in Fig. 6.8, was evolved in order to achieve tighter coupling than is possible with the edge-coupled arrangement of strips.

The analysis, and derivation of impedance formulae, has been carried out by Cohn [20, 21]: some of these formulae have already been encountered in Section 3.8.2, in connection with the High-Q triplate stripline. For the zero-thickness-strip case, conformal transformation techniques can be used to derive the following formulae, valid for all values of W/b, and h/b, *provided that* the ratio of these parameters exceeds 0·35, i.e. for $(W/b)/(h/b) > 0.35$,

$$Z_{0e}\sqrt{\kappa} = 59.952\pi \frac{K'(k)}{K(k)} \text{ ohm} \qquad (6.3.15)$$

$$Z_{0o}\sqrt{\kappa} = \frac{94.172\pi h/b}{\tanh^{-1}k} \text{ ohm} \qquad (6.3.16)$$

where

$$\frac{W}{b} = \frac{2}{\pi}\left\{\tanh^{-1}\sqrt{\frac{k(k-h/b)}{(1-kh/b)}} - \frac{h}{b}\tanh^{-1}\sqrt{\frac{(k-h/b)}{k(1-kh/b)}}\right\}$$

$$(6.3.17)$$

If, in addition, the cross-sectional dimensions are such that $W/b \geqslant 0.35 \,(1-h/b)$, the following simplified, and explicit, formulae can be used:

$$Z_{0e}\sqrt{\kappa} \doteqdot \frac{59.952\pi}{\{(W/b)/(1-h/b)\}+C_e} \text{ ohm} \qquad (6.3.18)$$

$$Z_{0o}\sqrt{\kappa} \doteqdot \frac{59.952\pi}{\{(W/b)/(1-h/b)\}+\{(W/b)/(h/b)\}-C_0} \text{ ohm} \qquad (6.3.19)$$

where

$$C_e = 0.4413 - H \qquad (6.3.20)$$

$$C_0 = Hb/h \qquad (6.3.21)$$

and

$$H = \frac{(1-h/b)\ln(1-h/b)+(h/b)\ln(h/b)}{\pi(1-h/b)}. \qquad (6.3.22)$$

The zero-thickness-strip case has also been treated by Yamamoto [22]. Using similar conformal mapping techniques, it is shown that Z_{0e} and Z_{0o} are both given by equation (6.3.15), but with different values of the modulus k for the even- and odd-mode cases. The implicit formulae relating k to the line parameters are rather involved, and for most design applications the use of Cohn's much simpler formulae (6.3.15)–(6.3.22) is recommended.

No exact treatment of the finite-thickness-strip case is known to the author, but Cohn [21] suggests the use of the zero-thickness formulae in conjunction with some correction formulae given in [23].

6.3.4 Offset Broadside-coupled Stripline

This type of coupled-strip transmission line shown in cross-section in Fig. 6.9 combines the advantages of the symmetrical broadside-coupled and edge-coupled structures already described, and provides a means of achieving a wide range of coupling values by the use of a fairly simple and easy-to-make configuration of strips and dielectric sheets.

An approximate analysis of the structure for

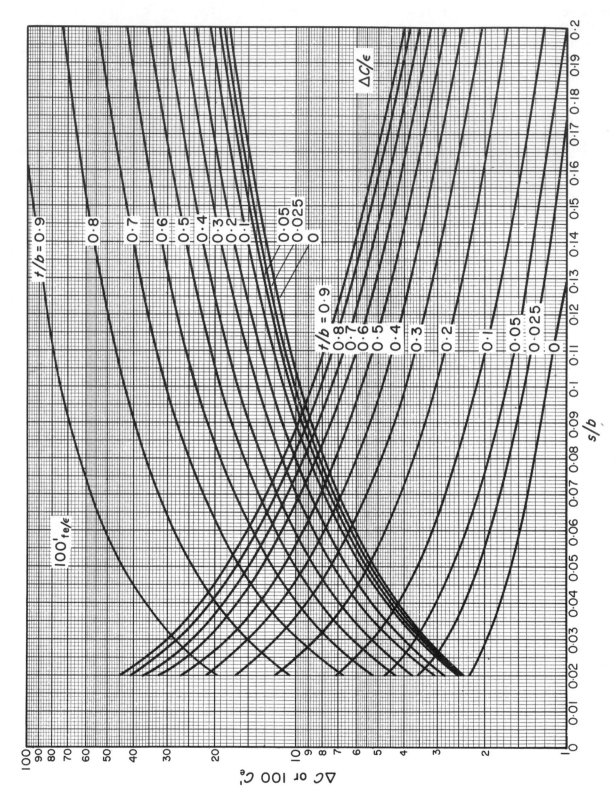

Fig. 6.6. Graphs of even-mode fringing capacitance C'_e, **and** $\Delta C = C'_0 - C'_e$ **(see Fig. 6.5 for definitions of parameters) for coupled finite-thickness triplate striplines, as calculated by Gupta [41] from Getsinger's formulae [17]. (By kind permission of the I.E.E.E.)**

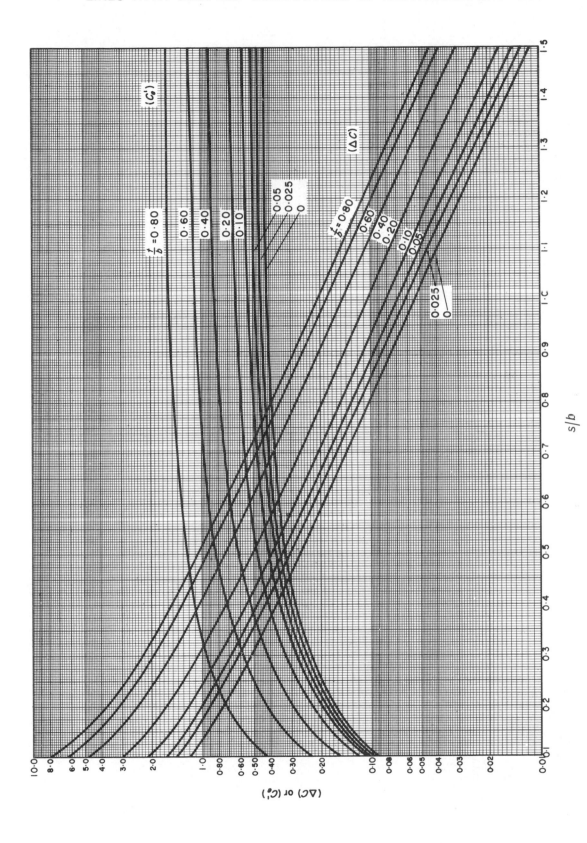

Fig. 6.7. (a) Graphs of even-mode fringing capacitance C_e', and $\Delta C = C_0' - C_e'$ for coupled finite-thickness triplate striplines. (Reproduced from [17] by kind permission of the I.E.E.E.)

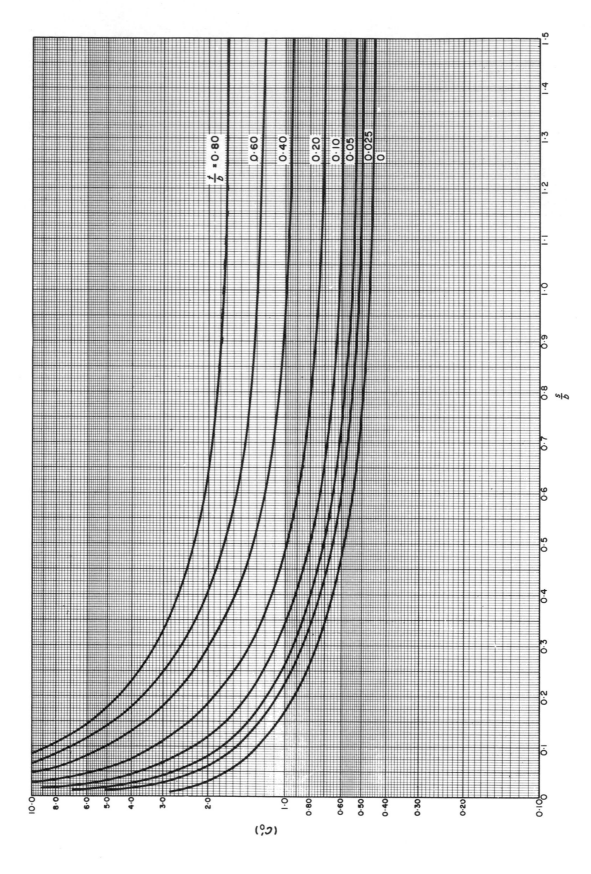

Fig. 6.7. (b) Graphs of odd-mode fringing capacitance C_0', for coupled finite-thickness triplate striplines. (Reproduced from [17] by kind permission of the I.E.E.E.)

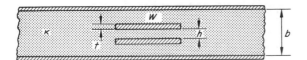

Fig. 6.8. Cross-section of broadside-coupled (horizontal) striplines

infinitely thin strips, was given by Shelton [24], using conformal transformation techniques to obtain formulae for the interelectrode capacitances, and hence the even- and odd-mode impedances. These formulae are rather lengthy, and since two different sets are required, to cater for the two cases of "tight" and "loose" coupling, they will not be given here. Instead, the reader/user may refer to Figs. 6.10(a) and 6.10(b), which display $Z_{0e}\sqrt{\kappa}$ and $Z_{0o}\sqrt{\kappa}$ as functions of the line dimensions. These were derived from computerized solution of Shelton's formulae. Note carefully that the following limitations are imposed by the approximations involved in the original analysis.

In general, results are valid provided that

$$\frac{W/b}{1-h/b} \geqslant 0.35$$

and further: for loose coupling,

$$\frac{2W_0/b}{1+s/b} \geqslant 0.85$$

For tight coupling,

$$\frac{W/b-W_0/b}{s/b} \geqslant 0.7.$$

Results (read from the graphs) which do not conform to these requirements should be used with caution.

6.3.5 Broadside-coupled (Vertical) Stripline

This form of coupled stripline (see Fig. 6.11) was originally proposed by Park [25], for use as a single transmission line, similar to the High-Q triplate line described in Section 3.8. It does not appear to offer any advantages over the other forms of strip transmission lines already dis-

cussed, and, indeed, suffers from two fairly obvious disadvantages: it is not as easy to fabricate, and is completely unsuitable for high-power operation because of the proximity of "sharp" edges to the ground-planes. However, it has found limited application in the design of group delay equalizers of "meander-line" construction [26, 27], for which a knowledge of Z_{0e} and Z_{0o} is required.

The analysis of this coupled-line structure, for zero-thickness strips, was given by Cohn, in the same paper [21] as the horizontally coupled version (see Section 6.3.3), and corrections for the finite thickness case were suggested. However, a more accurate treatment was given later [20], and the formulae derived (also quoted by Levy [10]) are as follows:

For zero-thickness strips ($t = 0$),

$$Z_{0e}\sqrt{\kappa} = 188.344 \frac{K(k)}{K'(k)} \text{ ohm} \qquad (6.3.23)$$

$$Z_{0o}\sqrt{\kappa} = \frac{295.849}{(b/s)\cos^{-1}k - \ln k} \text{ ohm.} \qquad (6.3.24)$$

For a given value of k, Z_{0e} and Z_{0o} are related to the line dimensions via

$$\frac{W}{b} = \frac{2}{\pi}\left(\tan^{-1}RX - \frac{s}{b}\tanh^{-1}X\right) \qquad (6.3.25)$$

where

$$R = k'/k, \quad k' = \sqrt{1-k^2} \qquad (6.3.26)$$

and

$$X = \sqrt{\frac{1-s/bR}{1+sR/b}}. \qquad (6.3.27)$$

These formulae are valid for all W/b and s/b, provided only that $W/s > 1$.

For finite-thickness strips ($t \neq 0$), the impedance formulae are modified as follows:

$$Z_{0e}\sqrt{\kappa} = \frac{188.344}{\{K(k)/K'(k)\} + \{(2t/b)/(1-W/b)\}} \text{ ohm} \qquad (6.3.28)$$

$$Z_{0o}\sqrt{\kappa} = \frac{94.172\pi}{\{(b/s)\cos^{-1}k - \ln k\} + \{(\pi t/b)/(1-W/b)\}} \text{ ohm} \qquad (6.3.29)$$

where k is related to W, s, t and b by the same formulae as before (equations (6.3.25)–(6.3.27)). Equations (6.3.28) and (6.3.29) yield accurate results provided that

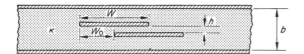

Fig. 6.9. Cross-section of offset broadside-coupled striplines

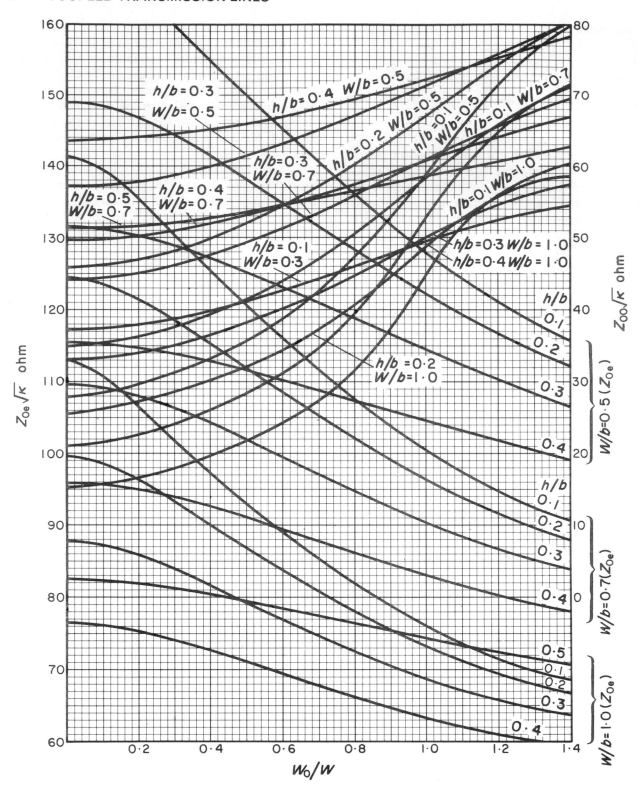

Fig. 6.10(a). Graph of $Z_{0e}\sqrt{\kappa}$ and $Z_0{}^o\sqrt{\kappa}$ for offset broadside-coupled striplines as calculated from Shelton's formulae [24]. See Fig. 6.9 for definition of parameters

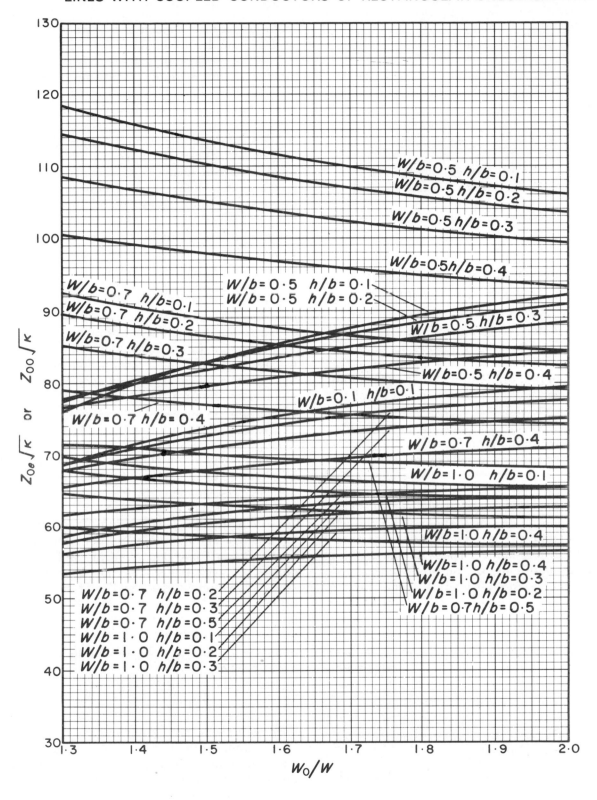

Fig. 6.10(b). Graph of $Z_{0e}\sqrt{\kappa}$ and $Z_{0o}\sqrt{\kappa}$ for offset broadside-coupled striplines as calculated from Shelton's formulae [24]. See Fig. 6.9 for definition of parameters

$$\frac{2t/b}{1-W/b} \ll 1$$

but results are in any case within a few per cent of "exact" values, for any value of t.

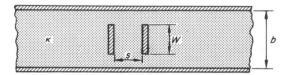

Fig. 6.11. Cross-section of broadside-coupled (vertical) striplines

6.3.6 Coupled High-Q Triplate Stripline

This structure (see Fig. 6.12) is fundamentally the same as that of two coupled pairs of the broadside-coupled (horizontal) striplines described in Section 6.3.3, except that the strips are supported by a plane layer of dielectric, of higher dielectric constant than that of the other medium filling the space between the ground-planes. As usual, the presence of two different dielectrics means that true TEM mode propagation is not possible, though, as with single-conductor duo-dielectric lines, the field structure closely resembles that which would be appropriate to a true TEM mode. This means that the usual electrostatic or pseudo-TEM techniques can be employed to calculate the even- and odd-mode impedances, but each will have associated with it a different value of effective dielectric constant κ_e. These can be denoted by κ_{ee} (even-mode) and κ_{eo} (odd-mode).

This is simply another way of saying that the two modes have different propagation velocities and wavelengths, which makes it difficult to define the quarter- and half-wavelength sections required in the design of filters and directional couplers. In practice, it is usual to take the arith-

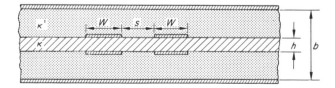

Fig. 6.12. Cross-section of coupled High-Q triplate striplines. Note that κ is the dielectric constant of "substrate", and κ' is the dielectric constant of medium filling the remaining space between the ground planes

metic or geometric mean of the two wavelengths in designing such sections.

The main effect of the unequal mode wavelengths is to reduce the bandwidth of operation from what it would have been with equal mode wavelengths. This is discussed in some detail by Levy [10].

Thus, although the High-Q triplate structure is simple and cheap to construct, the difficulties involved in its analysis have discouraged any intensive investigations, and consequently little design information is available. Cohn has considered the problem in [13], and suggests that the even-mode characteristic impedance Z_{0e} for coupled zero-thickness High-Q triplate will be approximately the same as for coupled finite-thickness edge-coupled triplate (as given, for example, by equation (6.3.9), putting $t = h$, and $\kappa = \kappa_{ee}$). Provided that $s/h \gg 10$, Z_{0o} can be similarly obtained from the appropriate triplate formulae; but for smaller values of s/h, results will be too inaccurate for design use, and to cover this situation Cohn gives a very lengthy and complicated formula involving the fringing capacitances and characteristic impedances of zero-thickness and finite-thickness edge-coupled triplate.

Other useful information can be obtained from an extensive analysis of multiconductor striplines published by Duncan [28]. He considers the four possible modes of excitation of two pairs of coupled strips between ground-planes, containing a single uniform dielectric: this corresponds to the "basic" line described in Section 3.8, in connection with the single High-Q triplate line. A variational analysis is used to obtain accurate values for the characteristic impedances of the four modes, and the relationships between Duncan's impedances and the High-Q triplate impedances Z_{0e} and Z_{0o} at present under discussion are as follows:

$$Z_{0e} = \tfrac{1}{4}Z_{04} \qquad (6.3.30)$$

$$Z_{0o} = Z_{03} \qquad (6.3.31)$$

Some fairly detailed graphs of $Z_{03}\sqrt{\kappa}$ and $Z_{04}\sqrt{\kappa}$ as functions of the line dimensional parameters are given in [28]. These can be used, in conjunction with equations (6.3.30) and (6.3.31), for design purposes, provided, of course, that suitable values for κ_{ee} and κ_{eo} can be obtained, either by preliminary measurement or from formulae. In the latter connection, an approximate formula given by Cohn [13] may be of some value: it is stated that in the High-Q triplate structure, with zero-thickness strips, the ratio of the odd- and even-mode velocities is given by:

$$\frac{v_0}{v_e} = \sqrt{\frac{Z^2 + 2Z_s Z_0}{Z^2 + 2\kappa Z_s Z_0}} \qquad (6.3.32)$$

where $Z = 376.687$ ohm (characteristic impedance of free space), Z_s is the odd-mode characteristic impedance for *edge-coupled zero-thickness triplate* (see Section 6.3.1) of width W, separation s, and ground-plane spacing b; Z_0 is the characteristic impedance of *single triplate* (see Section 3.5.2) of zero-thickness, but of width $W = s$, ground-plane spacing $b = h$, and with $\kappa = 1$; and κ is the dielectric constant of plane dielectric layer supporting the strips (see Fig. 6.12).

Since v_e is *approximately* the same as the velocity of light, particularly for small h, an approximate value for v_0 can be obtained from equation (6.3.28). Hence,

$$\kappa_{ee} \doteq \frac{c^2}{v_e^2} \doteq 1 \qquad (6.3.33)$$

$$\kappa_{eo} \doteq \frac{c^2}{v_0^2} \doteq \frac{v_e^2}{v_0^2}. \qquad (6.3.34)$$

6.3.7 Coupled Microstrip Line

Various forms of coupled microstrip lines are now widely employed in microwave integrated circuits, and the literature devoted to their design, analysis, and measurement is almost as extensive as that devoted to the single microstrip line. Additional stimulus, if any were required, has come from the area of high-speed computers, which, to an increasing extent, are employing single microstrip tracks in high-density assemblies of parallel circuits. Here, of course, the emphasis is on *elimination* of coupling between lines, in order to avoid "cross-talk" and similar undesirable effects, and the calculation of cross-coupling via the even- and odd-mode impedances is of great importance in determining the minimum track separation compatible with an acceptable level of cross-talk.

The basic coupled microstrip structure is illustrated in Fig. 6.13, and many approaches, both semi-analytic and numerical, have been employed in attempting to calculate its propagation parameters.

Schwarzmann [29, 30] has presented empirical closed-form formulae for Z_{0e} and Z_{0o}, but these

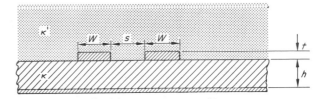

Fig. 6.13. Cross-section of coupled microstrip lines

involve "log" terms, with no indication of whether these are to base 10 or base "e": using either choice, results obtained in practice do not appear to be very accurate or reliable.

All other treatments known to the author involve computer application of numerical techniques; either directly, or to the solution of systems of equations derived by analytic means, and the majority of these use the "pseudo-TEM" mode assumption.

The data displayed in Figs. 6.14–6.17 were derived from a computerized extension of Silvester's analysis [31], as outlined in Section 3.6. Similar data, in tabulated form, have been presented by Bryant and Weiss [32], for zero-thickness strips on a number of different substrate materials: and by Hill *et al.* [33] for the case $\kappa = 4.4$ (appropriate to computer circuit interconnections).

Other methods of analysis of coupled microstrip, both shielded and unshielded, can be found in [34]–[40].

The important problem of the effects of dispersion, caused by the presence of two different dielectric media, has been considered by a number of investigators. To the author's knowledge, no exact modal analysis has yet been successfully completed, and most of the available data has been obtained by practical measurement. Deutsch and Jung [43] have carried out numerous measurements of κ_e, using alumina substrates in the frequency range from 2 to 12 GHz. They find variations of between 2 and 7 per cent from theoretically-calculated results (TEM mode assumption). Similar work and results have been reported by Napoli and Hughes [44], for the 4 GHz region.

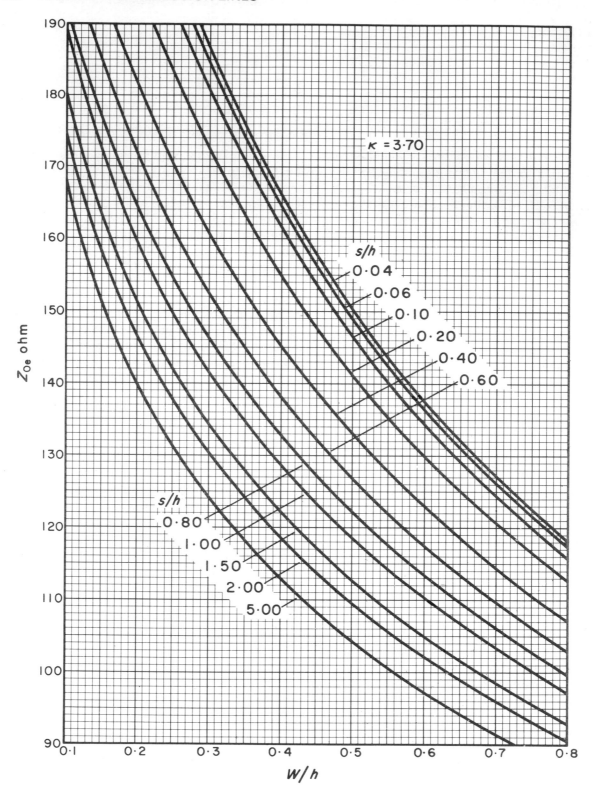

Fig. 6.14(a). Graph of even-mode and odd-mode characteristic impedances for zero-thickness coupled microstrip lines on a substrate of dielectric constant $\kappa = 3.70$ (e.g. quartz). See Fig. 6.13 for definition of parameters

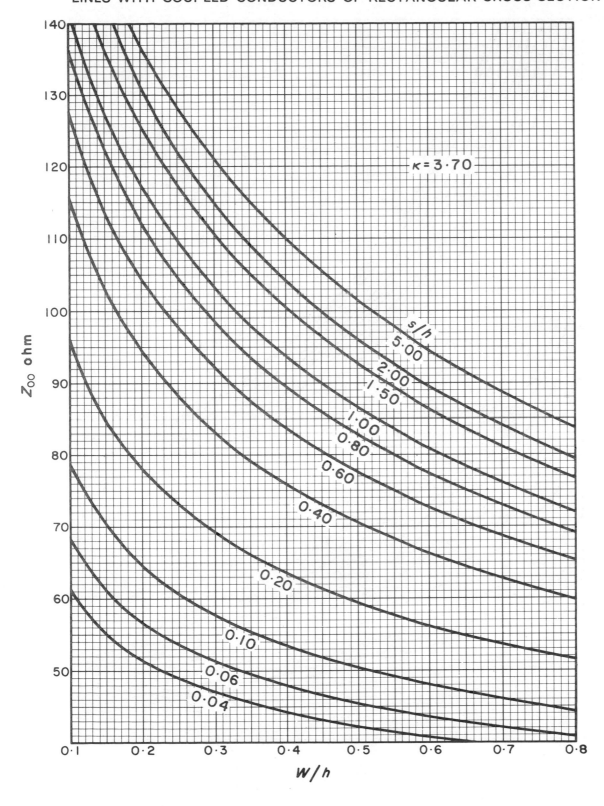

Fig. 6.14(b). Graph of even-mode and odd-mode characteristic impedances for zero-thickness coupled microstrip lines on a substrate of dielectric constant κ = 3·70 (e.g. quartz). See Fig. 6.13 for definition of parameters

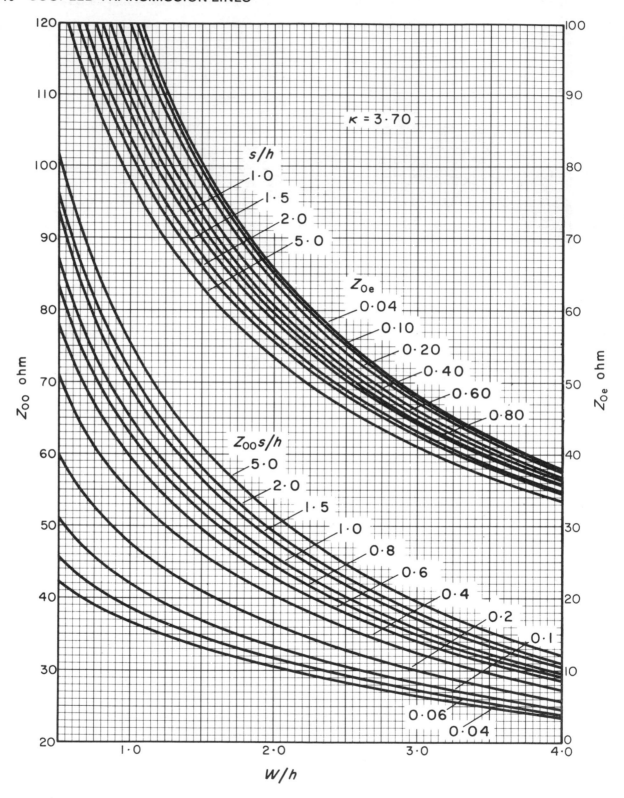

Fig. 6.14(c). Graph of even-mode and odd-mode characteristic impedances for zero-thickness coupled microstrip lines on a substrate of dielectric constant $\kappa = 3\cdot70$ (e.g. quartz). See Fig. 6.13 for definition of parameters

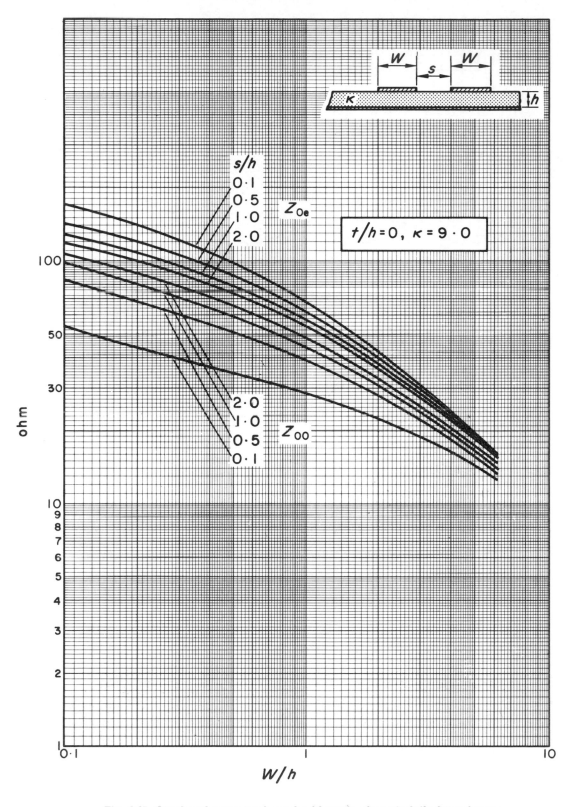

Fig. 6.15. Graphs of even-mode and odd-mode characteristic impedances
for zero-thickness coupled microstrip lines on a substrate of dielectric constant $\kappa = 9.0$ (e.g. alumina). See Fig. 6.13 for definition of parameters

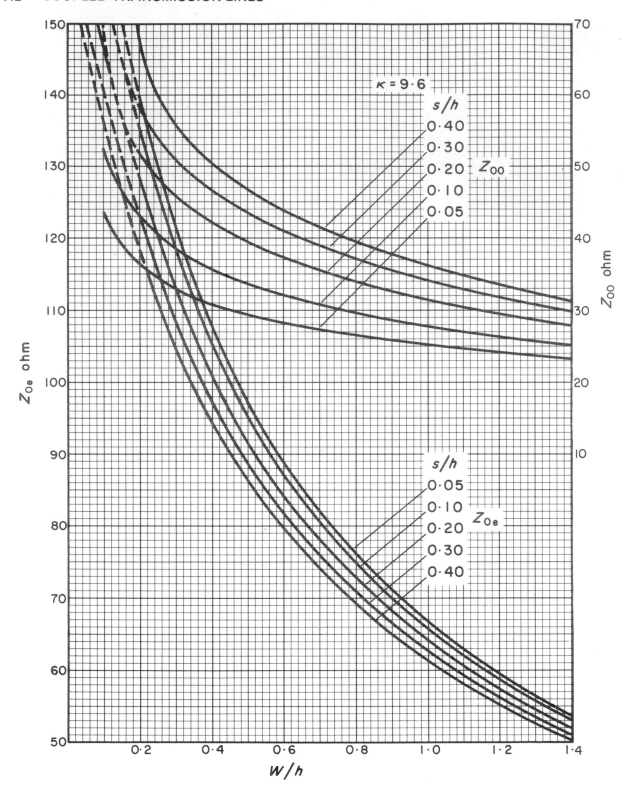

Fig. 6.16(a). Graph of even-mode and odd-mode characteristic impedances for zero-thickness coupled microstrip lines on a substrate of dielectric constant κ = 9·6 (e.g. alumina). See Fig. 6.13 for definition of parameters

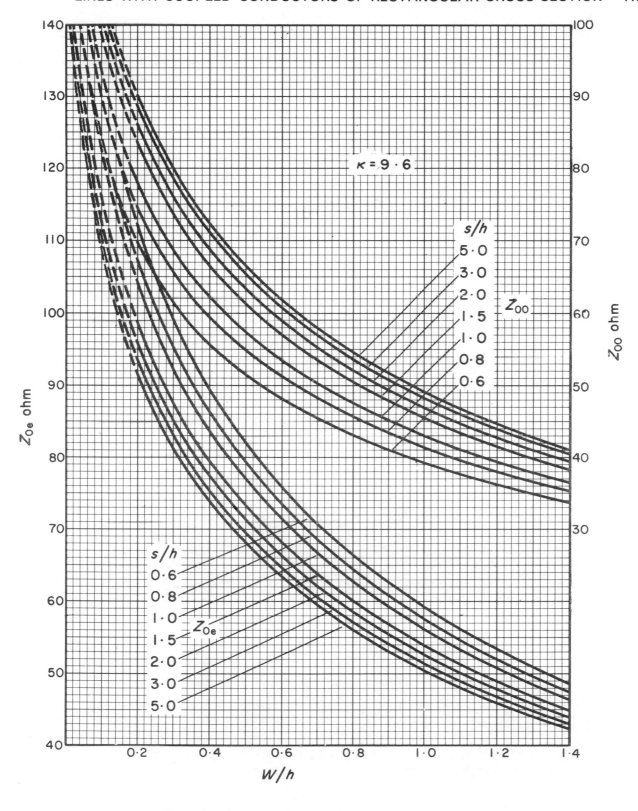

Fig. 6.16(b). Graph of even-mode and odd-mode characteristic impedances for zero-thickness coupled microstrip lines on a substrate of dielectric constant $\kappa = 9.6$ (e.g. alumina). See Fig. 6.13 for definition of parameters

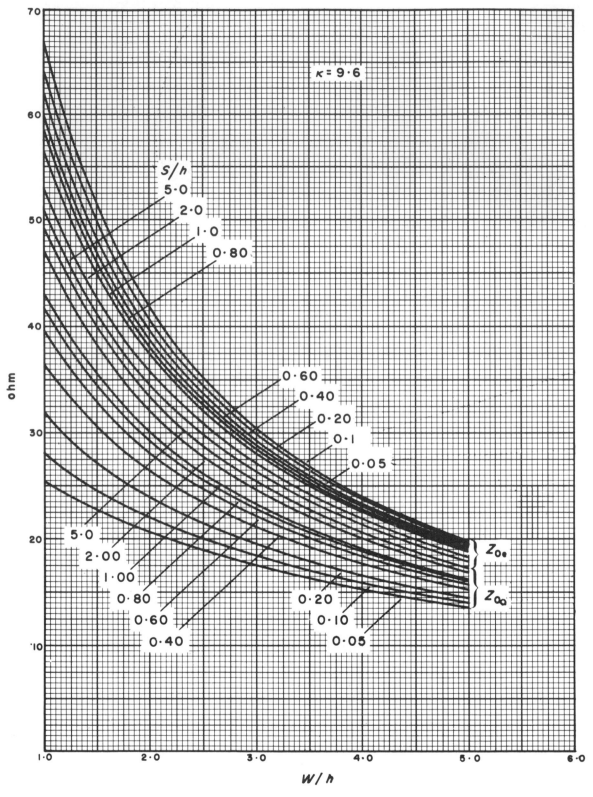

Fig. 6.16(c). Graph of even-mode and odd-mode characteristic impedances for zero-thickness coupled microstrip lines on a substrate of dielectric constant $\kappa = 9.6$ (e.g. alumina). See Fig. 6.13 for definition of parameters

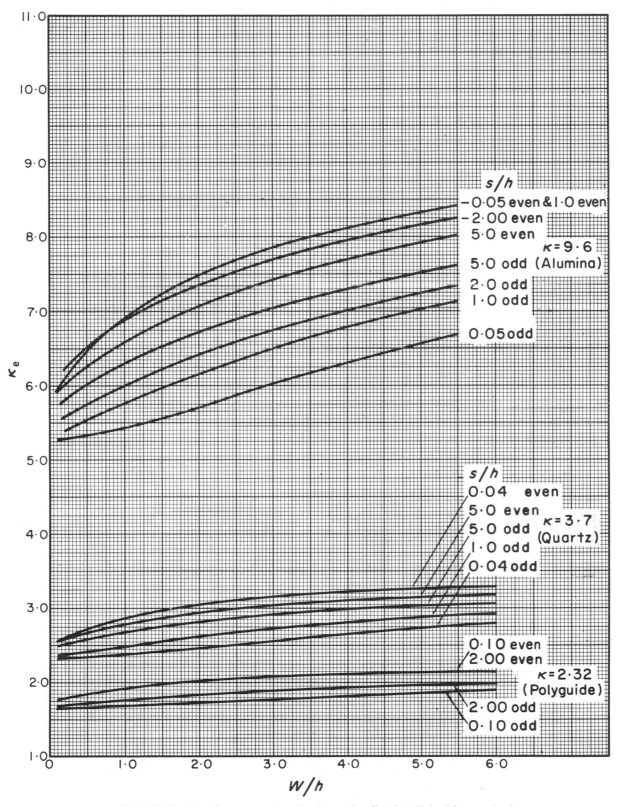

Fig. 6.17. Graphs of even-mode and odd-mode effective dielectric constants (κ_e) for zero-thickness coupled microstrip lines on substrates of dielectric constants $\kappa = 2.32$ (Polyguide), 3.70 (quartz), and 9.60 (alumina)

REFERENCES

1. Jones, E. M. T. and Bolljahn, J. T. "Coupled Strip Transmission Line Filters and Directional Couplers", *Trans. I.R.E.*
   MTT-4, p. 75 (1956).
2. Shimizu, J. K. and Jones, E. M. T. "Coupled Transmission Line Directional Couplers", Ibid.
   MTT-6, p. 403 (1958).
3. Gunston, M. A. R. and Nicholson, B. F. "Interdigital and Comb-Line Filters", *The Marconi Review.*
   3rd Qu., pp. 133–147 (1966).
4. Nicholson, B. F. "The Practical Design of Interdigital and Comb-Line Filters", *Radio and Electronic Engr.*
   34, pp. 39–52 (July, 1967).
5. Humphries, J. G. "Coaxial-Line Components for use at Microwave Frequencies", *I.E.E. Conf. Digest "Components for Microwave Circuits".*
   pp. 68–72 (Sept., 1962).
6. Frankel, S. "Characteristic Impedances of parallel wires in rectangular troughs", *Proc. I.R.E.*
   30, 182–190 (1942).
7. Bolljahn, J. T. and Matthaei, G. L. "A Study of the Phase and Filter Properties of Arrays of Parallel Conductors between Ground Planes", *Proc. I.R.E.*
   50, 299–311 (March, 1962).
8. Vadopalas, P. "Coupled Rods between Ground Planes", *Trans. I.E.E.E.*
   MTT-13, pp. 254–255 (March, 1965).
9. Cristal, E. G. "Coupled Circular Cylindrical Rods between Parallel Ground Planes", *Trans. I.E.E.E.*
   MTT-12, pp. 428–438 (July, 1964).
10. Levy, R. "Directional Couplers", *Advances in Microwaves.*
    Vol. I, pp. 115–209, Academic Press, New York (1966).
    See also Proc. I.E.E., **112**, 469–476 (March, 1965).
11. Conning, S. W. "The Capacitances of Cylindrical Bars between Parallel Ground Planes", *Proc. I.R.E.E. (Australia).*
    p. 131 (April, 1970).
12. Pon, C. Y. "A Wide-Band 3 dB Hybrid Using Semi-Circular Coupled Cross-Section", *Microwave J.*
    12, 81–85 (Oct., 1969).
13. Cohn, S. B. "Shielded Coupled-Strip Transmission Line", *Trans. I.R.E.*
    MTT-3, pp. 29–38 (Oct., 1955).
14. Singletary, J. "Fringing Capacitances in Strip-Line Coupler Design", *Trans. I.E.E.E.*
    MTT-14, p. 398 (Aug., 1966).
15. Sato, R. and Ikeda, T. "Line Constants".
    Chapter V in "Microwave Filters and Circuits" (Ed. A. Matsumoto). Supplement 1 to *Advances in Microwaves,* Academic Press, New York (1970).
16. Horgan, J. D. "Coupled strip transmission lines with rectangular inner conductors", *Trans. I.R.E.*
    MTT-5, pp. 92–99 (April, 1957).
17. Getsinger, W. J. "Coupled Rectangular Bars between Parallel Plates", *Trans. I.R.E.*
    MTT-10, pp. 65–72 (Jan., 1962).
18. Matthaei, G. L. *et al. Microwave Filters, Impedance-Matching Networks, and Coupling Structures.*
    pp. 194–195, McGraw-Hill, New York (1964).
19. Ilenburg, W-R. and Pregla, R. "Die Grenzfrequenzen von höheren Wellenformen in einer Anordnung mit mehreren Streifenleitungen", *A.E.Ü.*
    22, 230–238 (May, 1968).
20. Cohn, S. B. *et al.* "Design Criteria for Microwave Filters and Coupling Structures", *Tech. Rept. 2.*
    Stanford Research Institute Project 2326 (June, 1968).
21. Cohn, S. B. "Characteristic Impedances of Broadside-Coupled Strip Transmission Lines", *Trans. I.R.E.*
    MTT-8, pp. 633–637 (Nov., 1960).
22. Yamamoto, S. *et al.* "Slot-Coupled Strip Transmission Lines", *Trans. I.E.E.E.*
    MTT-14, pp. 542–553 (Nov., 1966).
23. Cohn, S. B. "Thickness corrections for capacitive obstacles and strip conductors", *Trans. I.R.E.*
    MTT-8, pp. 638–644 (Nov., 1960).
24. Shelton, J. P. "Impedances of Offset Parallel-Coupled Strip Transmission Lines", *Trans. I.E.E.E.*
    MTT-14, pp. 7–15 (Jan., 1966).
25. Park, D. "Planar transmission lines", *Trans. I.R.E.*
    MTT-3, pp. 8–12 (April, 1955); pp. 7–11 (Oct., 1955).
26. Pregla, R. "Delay Equalization with Transmission Line Circuits", *Proc. Colloquium on Microwave Communication.*
    pp. 315–325 (Budapest, 1968).
27. Tu, P. J. "A Computer-Aided Design of a Microwave Delay Equalizer", *Trans. I.E.E.E.*
    MTT-17, pp. 626–634 (Aug., 1969).
28. Duncan, J. W. "Characteristic Impedances of Multiconductor Strip Transmission Lines", *Trans. I.E.E.E.*
    MTT-13, pp. 107–118 (Jan., 1965).
29. Schwarzmann, A. "Microstrip plus equations adds up to fast designs", *Electronics.*
    pp. 109–112 (Oct. 2nd, 1967).
30. Schwarzmann, A. "Approximate solutions for a coupled pair of microstrip lines in microwave integrated circuits", *Microwave J.*
    12, pp. 79–82 (May, 1969).
31. Silvester, P. "TEM wave properties of microstrip transmission lines", *Proc. I.E.E.*
    115, pp. 43–48 (Jan., 1968).
32. Bryant, T. G. and Weiss, J. A. *Parameters of Microstrip Transmission Lines and of Coupled Pairs of Microstrip Lines.*
    Worcester Polytechnic Inst. Dept. of Physics (July 15th, 1968).
    See also same title in *Trans. I.E.E.E.*
    MTT-16, pp. 1021–1027 (Dec., 1968).
33. Hill, Y. M. *et al.* "A General Method for obtaining Impedance and Coupling Characteristics of Practical Microstrip and Triplate Transmission Line Configurations", *IBM Jl. Res. Dev.*
    pp. 314–322 (May, 1969).
34. Bräckelmann, W. "Wellentypen auf der Streifenleitung mit rechteckigem Schirm", *A.E.Ü.*
    21, pp. 641–648 (Dec., 1967).
35. Hellman, M. E. and Palocz, I. "The Effect of Neighbouring conductors on the Currents and Fields in Plane Parallel Transmission Line", *Trans. I.E.E.E.*
    MTT-17, pp. 254–259 (May, 1969).
36. Zysman, G. I. and Johnson, A. K. "Coupled Transmission Line Networks in an Inhomogeneous Dielectric Medium", *Trans. I.E.E.*
    MTT-17, pp. 753–759 (Oct., 1969).
37. Weeks, W. T. "Calculation of Coefficients of Capacitance of Multiconductor Transmission Lines in the Presence of a Dielectric Interface", *Trans. I.E.E.E.*
    MTT-18, pp. 35–43 (Jan., 1970).
38. Chen, W. H. "Even and Odd Mode Impedance of Coupled Pairs of Microstrip Lines", Ibid.
    pp. 55–57.
39. Judd, S. V. *et al.* "An Analytical Method for Calculating Microstrip Transmission Line Parameters", *Trans. I.E.E.E.*
    MTT-18, pp. 78–87 (Feb., 1970).
40. Krage, M. K. and Haddad, G. I. "Characteristics of Coupled Microstrip Transmission Lines", *Trans. I.E.E.E.*
    MTT-18, pp. 217–228 (April, 1970).
41. Gupta, R. R. "Fringing Capacitance Curves for Coplanar Rectangular Coupled Bars", *Trans. I.E.E.E.*
    MTT-17, pp. 637–638 (Aug., 1969).
42. Horton, M. C. "Loss Calculations for Rectangular Coupled Bars", *Trans. I.E.E.E.*
    MTT-18, pp. 736–738 (Oct., 1970).
43. Deutsch, J. and Jung, H. J. "Measurement of the Effective Dielectric Constant of Microstrip Lines in the Frequency Range from 2 GHz to 12 GHz", *NTZ.*
    23, 620–624 (Dec., 1970).
44. Napoli, L. S. and Hughes, J. J. "Characteristics of Coupled Microstrip Lines", *RCA Rev.*
    31, 479–498 (Sept., 1970).

INDEX